INFORM

Environmental Dividends
Cutting More Chemical Wastes

Mark H. Dorfman, Warren R. Muir, Ph.D.,
Catherine G. Miller, Ph.D.

INFORM, Inc.
381 Park Avenue South
New York, NY 10016-8806
Tel 212 689-4040
Fax 212 447-0689

TD
899
.C5
D66
1992

© INFORM, Inc. All rights reserved
Printed in the United States of America

Library of Congress Cataloging-in-Publication Data

Dorfman, Mark.
    Environmental dividends : cutting more chemical wastes/ Mark H. Dorfman, Warren R. Muir, Catherine G. Miller.
    p.   cm.
    Includes bibliographical references and index.
    ISBN 0-918780-50-0 : $75.00
    1. Chemical plants--United States--Waste disposal. 2. Hazardous wastes--United States. 3. Chemistry, Organic--United States. 4. Source reduction (Waste management)--United States. I. Muir, Warren R. II. Miller, Catherine G. III. Title.
TD899.C5D66   1992
660-dc20   92-12251
                                                                       CIP

INFORM, Inc., founded in 1974, is a nonprofit research organization that identifies and reports on practical actions for the protection and conservation of natural resources and public health. INFORM's research is published in books, abstracts, newsletters, and articles. Its work is supported by contributions from individuals and corporations and by grants from over 40 foundations.

Printed on recycled paper

TABLE OF CONTENTS

Part I

Chapter 1: Introduction 1
    The Dramatic Promise of Source Reduction 2
    INFORM's New Study: Evaluating Chemical Industry Progress 3
    Economic Climate of the US Chemical Industry 3
    Methodology of the Study 6
    Scope of the Study 10
Chapter 2: Findings and Conclusions 11
    Source Reduction Accomplishments at the Study Plants 14
    Program Features and Plant Characteristics 27
    Motivation 44
    Source Reduction Techniques 48
    Effect of Source Reduction on TRI Releases and Transfers 88
    Conclusions 89

Part II

Introduction to the Plant Profiles 91
American Cyanamid Company 94
Aristech 101
Atlantic Industries 110
Bonneau Dye Corporation 115
Borden Chemical Company 116
Chevron Chemical Company 123
Ciba-Geigy Corporation 128

Colloids of Califonia	141
Def-Tec Corporation	144
Dow Chemical USA	147
E. I. Du Pont De Nemours and Company	154
Exxon Chemical Americas	164
Fibrec, Inc.	171
Fisher Scientific Company	172
Frank Enterprises, Inc.	182
Hart Chem/J. E. Halma	183
ICI Americas, Inc.	184
ICI Resins	188
International Flavors and Fragrances, Inc.	193
Max Marx Color and Chemical Company	200
Merck and Company, Inc.	202
Monsanto Company	210
Morton International, Inc.	217
Perstorp Polyols, Inc.	220
PMC Specialities Group	224
Rhône-Poulenc, Inc.	229
Scher Chemicals, Inc.	236
Shell Chemical Company	240
Unocal Chemicals	243

Appendices

A: Statistical Tests Performed on Program Feature Data	247
B: Methodology for Estimating Impact of Source Reduction on Waste Generation and TRI Releases and Transfers	249
C: Bibliography	252
D: Glossary	253
Index	259

TABLES

Part I

Table I-1:	INFORM's Study Plants	4
Table I-2:	Number of Source Reduction Activities for Which Plants Provided Information, by Category	10
Table I-3:	Summary of Source Reduction Data by Plant	12
Table I-4:	Source Reduction Activities in Production and Nonproduction Functions	24
Table I-5:	Source Reduction Program Features at INFORM Study Plants	28
Table I-6:	Average Number of Source Reduction Activities per Plant by Program Feature	31
Table I-7:	Plant Characteristics and Number of Source Reduction Activities	34
Table I-8:	Source Reduction Program Features at the INFORM Study Plants	36
Table I-9:	Factors Motivating Source Reduction Activities	44
Table I-10:	Number of Source Reduction Activities by Motivating Factor and Type of Waste	47
Table I-11:	Average Percent Wastestream Reduction by Source Reduction Technique Used	50
Table I-12:	Number of Source Reduction Activities by Type of Waste and Source Reduction Technique	50
Table I-13:	137 Source Reduction Activities Categorized by Technique Used	52

Part II

Table II-1:	American Cyanamid: Source Reduction Activities	98
Table II-2:	Aristech: Source Reduction Activities	104
Table II-3:	Aristech: Changes in Waste Generation, 1984-1987	107
Table II-4:	Atlantic Industries: Source Reduction Activities	112

Table II-5:	Borden Chemical Company: Source Reduction Activities	120
Table II-6:	Chevron Chemical Company: Source Reduction Activities	126
Table II-7:	Ciba-Geigy: Source Reduction Activities	134
Table II-8:	Ciba-Geigy: Wastewater Effluent, 1979-1987	138
Table II-9:	Colloids of California: Source Reduction Activities	142
Table II-10:	Def-Tec Corporation: Source Reduction Activities	144
Table II-11:	Dow Chemical USA: Change in RCRA Hazardous Wastes, 1984-1988	150
Table II-12:	Dow Chemical USA: Source Reduction Activities	152
Table II-13:	Du Pont: Source Reduction Activities	160
Table II-14:	Exxon Chemical Americas: Source Reduction Activities	168
Table II-15:	Fisher Scientific: Increases in Overall Product Yields for Two Solvent Distillation Operations	174
Table II-16:	Fisher Scientific: Source Reduction Activities	176
Table II-17:	ICI Americas: RCRA Waste Generation, 1981-1988	186
Table II-18:	ICI Americas: TRI Toxic Chemical Releases, 1987 and 1988	186
Table II-19:	ICI Americas: Source Reduction Activities	187
Table II-20:	ICI Resins: Source Reduction Activities	190
Table II-21:	IFF: RCRA Hazardous Waste Generation, 1984-1987	194
Table II-22:	IFF: TRI Toxic Chemical Releases, 1987 and 1988	195
Table II-23:	IFF: Changes in Selected Wastes, 1986-1987	196
Table II-24:	IFF: Source Reduction Activities	198
Table II-25:	Merck: Chemical Input and Waste Generation, 1978-1988	204
Table II-26:	Merck: Toxics Release Inventory Releases, 1987 and 1988	206
Table II-27:	Merck: Source Reduction Activities	208
Table II-28:	Monsanto: Source Reduction Activities	212
Table II-29:	Morton International: RCRA Hazardous Waste, 1984-1989	218
Table II-30:	Morton International: TRI Toxic Chemical Releases, 1987-1989	218
Table II-31:	Perstorp Polyols: Source Reduction Activities	222
Table II-32:	PMC Specialities Group: Source Reduction Activities	226
Table II-33:	Rhône-Poulenc: Source Reduction Activities	232
Table II-34:	Scher Chemicals: Source Reduction Activities	238
Table II-35:	Shell Chemical: TRI Toxic Chemical Releases and Transfers, 1987 and 1988	241
Table II-36:	Shell Chemical: Generation of RCRA Hazardous Wastes, 1985-1987	242

FIGURES

Figure 1: Source Reduction Activities by Year First Implemented 14

Figure 2: Waste Reduced per Source Reduction Activity 15

Figure 3: Percentage of Target Wastestream Reduced per Source Reduction Activity 15

Figure 4: Cumulative Number of Source Reduction Activities per Year,
by Type of Waste 16

Figure 5: Average Percent Reduction per Source Reduction Activity
by Type of Waste 17

Figure 6: Implementation Time per Source Reduction Activity 18

Figure 7: Percent Increase in Product Yield per Source Reduction Activity 19

Figure 8: Annual Savings per Source Reduction Activity 20

Figure 9: Capital Costs per Source Reduction Activity 21

Figure 10: Average Payback Period per Source Reduction Activity by Range
of Capital Costs 21

Figure 11: Payback Periods per Source Reduction Activity 22

Figure 12: Annual Dollar Savings per Dollar of Capital Investment per
Source Reduction Activity 23

Figure 13: Source Reduction Activities in Production and Nonproduction Functions 26

Figure 14: Average Number of Source Reduction Activities by Number of Source
Reduction Program Features Established per Plant 30

Figure 15: Number of Plants Fully or Partially Establishing Source Reduction
Program Features 30

Figure 16: Motivation for Source Reduction Activities by Year 45

Figure 17: Motivation for Implementing Source Reduction Activities by Type
of Waste 47

Figure 18: Cumulative Number of Source Reduction Activities per Year by
Source Reduction Technique Used 49

Figure 19: Techniques Used for Source Reduction Activities by Type of Waste 51

ACKNOWLEDGMENTS

We wish to thank many people for their valuable contributions to this report. The nature of this study made it most crucial to have the cooperation of officials at the 29 chemical manufacturing facilities we studied. We therefore would like to express our appreciation to those plant officials who responded to our request for on-site interviews with invitations for plant visits. These officials often provided us with detailed responses to questionnaires sent prior to the visit and then took time to review and update the profiles we had prepared. We also thank officials from other plants who provided some information by phone or mail.

We extend special thanks to David J. Sarokin, coauthor of INFORM's 1985 study, *Cutting Chemical Wastes*, and currently Environmental Protection Specialist at the US Environmental Protection Agency, for his insight, suggestions, and reviews. We also thank Philip J. Landrigan, M.D., of the Division of Environmental and Occupational Medicine at New York's Mount Sinai Medical Center, for his helpful comments on a draft of this study.

Many people at INFORM contributed to the successful completion of this report. We thank Joanna D. Underwood, President of INFORM, and Nancy Lilienthal, Director of INFORM's Chemical Hazards Prevention Program, for their support throughout research, writing, and production. Special thanks also go to Sibyl R. Golden, Director of Research and Publications, for her expert guidance and tireless energy in helping create a document that is thorough, concise, and clear.

Eric J. Dolin carried out the earliest stages of research for this project, and Michèle Ascione provided research assistance in its later stages. Thanks also to Wally Wentworth, Mia Fienemann, and Ellen Lubell for their comments and suggestions.

An award for dedication to eye-straining details goes to Elisa Last, Production Coordinator, and Diana Weyne, Editorial Assistant, for their meticulous and creative production and editing efforts.

Finally, we thank the many individuals and organizations that provided general support, and the following foundations whose grants made this study possible: The Mary Reynolds Babcock Foundation, the Geraldine R. Dodge Foundation, The Fund for New Jersey, The George Gund Foundation, the W. Alton Jones Foundation, Inc., The Joyce Foundation, The S. Livingston Mather Charitable Trust, the Charles Stewart Mott Foundation, Inc., The Sears Family Foundation, and the Victoria Foundation, Inc.

While we could not have gathered and analyzed the information in this book without the assistance of all these people, the findings and conclusions are the sole responsiblity of INFORM.

PREFACE

The last decade has seen a virtual revolution in thinking in the United States and around the world about ways to confront mounting global waste crises. In 1982, when we launched INFORM's study of industrial chemical wastes, the spotlight was on management and disposal of such wastes—strategies with soaring costs and plunging effectiveness. Pollution prevention, the notion of reducing wastes at their source, was merely an idea optimistically whispered about. Today, 10 years later, it stands front and center stage—a clear call to an environmentally sound future.

In 1985, INFORM research, published in our report, *Cutting Chemical Wastes*, defined for the first time, through plant case studies, the specific economic and environmental benefits of industrial source reduction — preventing the generation of wastes in the first place rather than cleaning them up after they have been created. Advantages of this preventive approach were discovered to include cost savings, improved product yield and quality, reduced pollution, safer workplace conditions, fewer waste management needs, and conservation of natural resources.

Cutting Chemical Wastes played an important role in bringing industrial source reduction, the measures that could accomplish it, and the vital need for public information about toxic and hazardous chemical wastes into the national spotlight. Since then, a plethora of significant changes have taken place.

- At the end of 1986, as part of the reauthorized Superfund law, Congress required the more than 20,000 manufacturing plants using the largest amounts of some 320 toxic chemicals to report their releases and transfers of these chemicals to all environmental media — air, land, and water. The picture this produced, in the now annual Toxics Release Inventory, of billions of pounds of wastes entering our environment startled the public and private sectors alike.

- In 1989, the US Environmental Protection Agency established its first Office of Pollution Prevention.

- A year later, the US Congress passed the first federal pollution prevention law, lending legislative impetus to EPA's program, creating a pollution prevention clearing house, providing funds for state pollution prevention programs, and expanding industry reporting requirements.

- The EPA's 33/50 program, announced in 1991, called on major companies to voluntarily cut generation of 17 high-priority chemical wastes in half by 1995.

- By early 1992, some 25 states had, on their own, adopted some form of pollution prevention initiative.

- Industry has taken action as well. The Chemical Manufacturers Association's Responsible Care Program includes public disclosure and pollution prevention as two of its basic precepts. Furthermore, by the time this report went to press, hundreds of CMA member companies had responded to EPA's voluntary 33/50 program, committing to cut their generation of the targeted 17 chemical wastes in half by 1995.

During this period of change, we decided to take a fresh look at the 29 organic chemical manufacturing plants profiled in *Cutting Chemical Wastes* to see what new source reduction activities they had undertaken, what economic and environmental benefits they were achieving, what was motivating them, and what techniques and program features were of most value in reducing wastes at source.

This report, *Environmental Dividends: Cutting More Chemical Wastes,* presents the exciting results of our research. From the few and far between pollution prevention activities reported in *Cutting Chemical Wastes*, significant steps forward have been taken. We have found the facilities studied to be identifying more and more opportunities for significantly reducing their wastes. Through analysis of the 181 individual source reduction initiatives reported to INFORM, we have been able to quantify, in much greater detail, cost savings, payback times, increased production efficiency, yield increases, and waste decreases — many attained through simple, low-cost or no-cost strategies, making source reduction a key to economic competitiveness.

The 29 chemical plants that we have studied are clearly just the tip of the iceberg. Tens of thousands of waste-generating facilities in the United States and in the growing chemical industries around the world — from Latin America, to Eastern Europe, to Asia — can realize the same kinds of economic and environmental benefits by searching for their own source reduction opportunities. We have come far in the last decade. However, *Environmental Dividends* points the way to the even greater gains that can — and must — be made in the decade ahead.

Joanna D. Underwood
President
INFORM

PART I   CHAPTER 1

Introduction

One of the great economic and environmental challenges facing the United States and, increasingly, other industrialized countries is the mounting quantity of hazardous and toxic wastes that are the by-products of chemical industry operations. The great contribution this industry's products have made to modern life — in the form of plastics, solvents, adhesives, pharmaceuticals, and much more — has come with a heavy price, one we are only beginning to understand.

Since 1987, the Environmental Protection Agency's Toxics Release Inventory (TRI) has provided detailed evidence of the vast amounts of toxic waste released into this country's air, land, and water and transferred to off-site treatment and disposal facilities each year. According to the most recent TRI report, the 22,650 largest US industrial facilities using toxic chemicals released or transferred 5.7 billion pounds of 322 toxic chemicals and chemical categories into our environment in 1989. Furthermore, the United States has more than 20,000 officially designated hazardous waste sites, and another 200,000 "unofficial" pits, ponds, and lagoons containing wastes suspected of being hazardous.

Yet the Toxics Release Inventory understates the total amount of chemical pollution actually generated in the United States. The 322 chemicals and chemical categories that were reported in 1989 represent only a fraction of the 70,000 chemicals currently produced for commercial use. Nonmanufacturing sources of these toxic chemicals, such as waste treatment plants, public utilities, or farms, are not covered. Federal, state, and local government facilities are not included. Finally, small manufacturing facilities and ones that generate wastes below specified threshold levels are not required to report. Despite these limitations, however, the Toxics Release Inventory provides the first systematic nationwide look at patterns of industrial toxic pollution. (For an explanation of INFORM's use of the terms "toxic and hazardous substances" in this study, see Box 1.)

The chemical and allied products industry — one of 20 industries required to report TRI data — has consistently generated nearly half of all TRI wastes. In 1989, 4,259 chemical facilities (19 percent of all plants reporting) released or transferred 2.7 billion pounds of TRI wastes (48 percent of the total). These chemical facilities reported an average of five individual chemicals in use at each facility, with at least one plant reporting on 85 separate chemicals.[1]

Toxic and hazardous wastes are also a great economic liability, threatening the competitiveness of United States industry. The cost of federally mandated pollution control and cleanup programs in the country was $100 billion in 1990, up from just $26 billion in 1972. Almost two-thirds of this expenditure is borne by industry, the rest by government.[2]

---

1 US Environmental Protection Agency, *Toxics in the Community: National and Local Perspectives: The 1989 Toxics Release Inventory National Report,* Washington DC, September 1991.
2 US Environmental Protection Agency, cited in *The New York Times,* December 23, 1990.

Box 1: Terminology Used in This Study

For the purposes of this study, INFORM uses the term "toxic or hazardous substances" to mean materials included by the federal government on any of six lists:

- Clean Air Act hazardous air emissions
- Clean Water Act priority pollutants
- Emergency Planning and Community Right-to-Know Act (also known as Title III of the Superfund Amendments and Reauthorization Act, or SARA) Section 313 toxic substances; this is the list used by the EPA for the Toxics Release Inventory
- Emergency Planning and Community Right-to-Know Act (also known as Title III of the Superfund Amendments and Reauthorization Act, or SARA) Section 302 extremely hazardous substances
- Resource Conservation and Recovery Act (RCRA) Appendix VIII: hazardous constituents
- Resource Conservation and Recovery Act (RCRA) commercial chemical products: U (hazardous wastes) and P (acute hazardous wastes)

The term "toxic or hazardous waste" is easily subject to misinterpretation because the words mean different things in different contexts. INFORM's use of the phrase is broader than the conventional uses of the terms which refer only to materials that companies must report to the Toxics Release Inventory (toxic substances) and to solid hazardous wastes regulated under RCRA (hazardous wastes). INFORM considers any type of wastestream hazardous or toxic if it contains one of the substances appearing on any of these six lists, whether or not the substance is regulated in the medium in which it is released, and regardless of its concentration.

In one sense, INFORM's use of the phrase "hazardous wastes" is more narrow than the regulatory usage, because RCRA includes in its list of hazardous wastes those substances characterized as flammable, corrosive, or explosive, as well as those that are toxic. INFORM has excluded substances that are solely flammable, corrosive, or explosive, but not toxic under RCRA.

The term "toxic or hazardous waste" as used in this study is not intended to imply that the concentration and amount of the chemicals in the various wastestreams are always present in quantities sufficient to cause harm. The focus of this study is on the particular chemical substances generated as wastes and does not include an analysis of the extent or severity of environmental or human exposures to these substances.

The term "wastes," as used in this report, includes pollutant discharges, off specification products, noncommercial co-products or by-products, and substances slated to receive destructive or containment treatments.

The Dramatic Promise of Source Reduction

In 1985, INFORM's study *Cutting Chemical Wastes* played an important role in bringing the strategy of industrial source reduction — preventing the generation of waste rather than cleaning it up after it has been created — into the national spotlight. This 535-page examination of 29 chemical plants defined for the first time the specific plant-level environmental and economic benefits of source reduction. In the same year, the congressional Office of Technology Assessment (OTA) estimated (in a study entitled *Serious Reduction of Hazardous Waste*) that industry could cut waste production nearly in half over 5 years; OTA has also estimated that industry could save $50 for every $1 spent by government on source reduction.

Cutting Chemical Wastes documented the fact that each one of the 13 study plants that looked for ways to reduce waste at source found important and exciting opportunities not only to significantly decrease the amounts of waste it was generating but also to realize considerable cost savings after relatively short payback periods. These plants reported undertaking a total of 44 initiatives — some resulted in virtual elimination of wastestreams, others in waste reductions of 80 percent or more. Almost all of these initiatives involved simple operations or equipment changes.

Over the last few years, an unusually strong national consensus has emerged on the value of source reduction. It has been endorsed by groups ranging from the Chemical Manufacturers Association to the Environmental Protection Agency, from state legislatures to environmental and community groups. In the closing days of the 1990 session, Congress passed the Pollution Prevention Act, its first serious action embracing source reduction as national environmental policy rather than focusing primarily on waste management. Despite this wide agreement on the joint environmental and economic benefits of source reduction, companies have been slow to move in this vital direction.

INFORM's New Study: Evaluating Chemical Industry Progress

INFORM undertook this new study of source reduction activities at the 29 organic chemical manufacturing plants originally profiled in *Cutting Chemical Wastes* in 1985 in order to assess the industry's source reduction progress. Given the increase in government, business, and community interest in the strategy, were the 29 plants implementing new source reduction activities? What economic and environmental benefits were they achieving? What motivated them to look for source reduction opportunities? What techniques and program features were used most often to reduce wastes at source?

This report draws on data collected in two rounds of research at the 29 plants. Information originally collected for *Cutting Chemical Wastes* describes 44 source reduction activities (at 13 plants) that were implemented before 1985. The new research that began in 1987 includes information on 137 additional source reduction activities (at 24 plants) implemented through 1990.[3]

Economic Climate of the US Chemical Industry

The overall economic picture of the US chemical industry is generally cyclical. It showed marked improvement between the period in which INFORM conducted research for the original *Cutting Chemical Wastes* report (1979-1984) and the period in which INFORM conducted its second round of research (1985-1990), with average annual sales, earnings, and profit margins (after-tax earnings as a percent of sales) all increasing. Annual sales increased from an average of about $55 billion per year to about $75 billion per year between these two time periods. Average annual earnings doubled from about $2.5 billion per year during the period 1979 to 1984 to $5 billion per year in the period from 1985 through 1990. Profit margins grew from an average of about 5 percent of sales in the earlier period to an average of about 7 percent in the latter period.[4] Some chemical industry leaders cite reduced operations costs as one of the major reasons for the success. Others cite "a further shucking of marginally profitable businesses... [and introduction of] new product lines."[5] In 1988, the successes were predominantly due to "higher prices and production."[6]

Although the average figures for the 1985-1990 period showed growth when compared to the earlier period, the picture began to change at the end of the period. In 1989, "rising feedstock costs and falling product prices"[7] began to erode the enormous gains made in the two previous years. Although selling prices increased total sales figures in 1990, "higher feedstock prices brought on by the crisis in the Middle East, and a continued soft economy"[8] resulted in continued erosion in earnings and profit margins for that year. And in 1991, sales, earnings, and profit margins all dropped, with *Chemical & Engineering News* citing the economic recession as the major cause.[9]

---

3 Data for 1990 do not include source reduction activities implemented in the last 2 months of the year.
4 *Chemical & Engineering News*, February 18, 1991.
5 *Chemical & Engineering News*, February 22, 1988.
6 *Chemical & Engineering News*, February 20, 1989.
7 *Chemical & Engineering News*, February 19, 1990.
8 *Chemical & Engineering News*, February 18, 1991.
9 *Chemical & Engineering News*, February 17, 1992.

Table I-1: INFORM's Study Plants

	State	Number of Employees (Original/Update)	Change in Production between Original and Update	Reported Source Reduction (Original/Update)	Cooperation with Study (Original/Update)
Large Plants (more than 100 employees)					
Aristech (formerly USS Chemicals)	OH	224/257	+10 to 200%	Yes/Yes	Yes/Yes
Atlantic	NJ	200/240	+15 to 0%	Yes/Yes	Yes/Yes
Chevron	CA	455/274	−22%	No/Yes	No/Partial
Ciba-Geigy	NJ	1,050/400	−50%	Yes/Yes	Yes/Yes
Dow	CA	650/715	+260%	Yes/Yes	Yes/Yes
Du Pont	NJ	4,100/3,500	0	No/Yes	No/Yes
Exxon	NJ	550/500		Yes/Yes	Yes/Yes
Fisher	NJ	325/130	0	No/Yes	Yes/Yes
ICI Americas (formerly Stauffer)	CA	500/		Yes/Yes	Yes/Partial
IFF	NJ	375/225	+62%	No/Yes	No/Partial
Merck	NJ	500/200		Yes/Yes	Yes/Partial
Monsanto	OH	800/850		Yes/Yes	Yes/Yes
Morton International (formerly Carstab)	OH	250/		No/No	No/Partial
PMC (formerly Sherwin-Williams)	OH	200/200	0	Yes/Yes	No/Yes
Medium Plants (50 - 100 employees)					
American Cyanamid	OH	128/ 90	+200%	No/Yes	No/Yes
Borden	CA	50/50	0	Yes/Yes	Yes/Yes
Def-Tec (formerly Smith & Wesson)	OH	100/ 65		No/Yes	No/Yes
Rhône-Poulenc	NJ	140/ 60	−19.5%	No/Yes	No/Yes
Shell	CA	59/		No/No	No/No
Unocal	CA	100/	+15 to 20%	Yes/No	Yes/Yes
Small Plants (fewer than 50 employees)					
Bonneau Dye	OH	2/		No/No	No/No
Colloids	CA	5/4	0	No/Yes	No/Yes
Fibrec (plant no longer exists)	CA	3/0	NA	No/NA	No/NA
Frank Enterprises (no longer in manufacturing)	OH	5/	NA	No/NA	No/No
Hart Chem/J. E. Halma	NJ	7/5		No/No	No/No
ICI Resins (formerly Polyvinyl)	CA	18/15		No/Yes	No/Yes
Max Marx	NJ	25/20	0	No/No	No/Yes
Perstorp Polyols	OH	37/37	+10 to 15%	Yes/Yes	Yes/Yes
Scher	NJ	31/20	+200%	No/Yes	Yes/Yes

Blank, information not supplied by officials at the plant.
NA, information not applicable.

Major Products

Phenol, acetone, alpha-methylstyrene (AMS), cumene hydroperoxide, bisphenol-A

Dyestuff

Agricultural chemicals including pesticides, herbicides, fungicides, and fertilizers

Dyes, epoxy resins, and additives (all production ended — now only formulating, blending, and repackaging of imported dyes)

Sodium hydroxide, chlorine, hydrogen, perchloroethylene, carbon tetrachloride, hydrochloric acid, chlorinated pyridines, latexes, and sulfuryl fluoride. There is also a large research and pilot plant testing facility at this site.

Chemicals for textiles, automobiles, agriculture, building industry, soaps and detergents, and intermediates. There is also a research facility at this site.

Chemical raw materials (olefins such as propylene, butylenes), additives for fuels and lubricating oils, synthetic lubricating oils, dispersant additives, specialty chemicals (such as isobutylene polymers for chewing gum and surgical adhesives)

Reagent chemicals

Agricultural chemicals

Flavors and fragrances for the food, beverage, cosmetics, and soap and detergent industries

Pharmaceuticals

Acrylonitrile-butadiene-styrene, styrene-acrylonitrile, polystyrene plastics, and styrene-maleic anhydride resin

Organotin PVC heat stabilizers, antioxidants, synthetic lubricants, asphalt additives, phosphonium salt polymerization catalysts, and extreme pressure lubricant additives

Saccharin, corrosion inhibitors, intermediates, additives, and specialty chemicals (isoatoic anhydride, anthranilic acid)

Vulcanized vegetable oils, ultraviolet absorbers, and organic intermediates

Formaldehyde, urea-formaldehyde and phenol-formaldehyde resins, adhesives, and wax emulsions

Tear gas production and light machine work and assembly of riot control equipment related to tear gas use

Aroma chemicals, rare earth compounds, and chemical intermediates

Inorganic metallic catalysts for dehydrogenation and hydrotreating processes

Latex polymers (polyvinyl acetate, styrene-butadiene, acrylic resins) and solvent blending (organic chemical solvents primarily for paint companies and janitorial supply companies — methanol, methyl ethyl ketone (MEK), methyl isobutyl ketone (MIK), dichloromethane, 1,1,1-trichloroethane, acetone, cyclohexane, textile solvents and rubber solvent)

Custom blend dye formulations and specialty chemicals for candle-making and other crafts

Simple blending and compounding of various liquids and dry materials for use as industrial antifoam agents; repackages additives and polyacrylates

No longer in business

Sales office only since explosion in 1986

Solvents, etchants, acids, and cleaners used in the semiconductor and transistor manufacturing industries

Acrylic polymers (for protective coating, floor care, and cement admixture industries)

Organic pigments (printing inks and paints)

Pentaerythritol (used in manufacture of paints, printing inks, and synthetic lubricants) and sodium formate (leather, textile, paper, and chemical industries)

Specialty chemicals for use in cosmetics and textile industries (acrylonitrile, maleic anhydride)

Between the two rounds of INFORM research, many of the INFORM study plants streamlined operations even though production was growing. Officials at 8 of the 17 plants that reported production output and employment figures for both time periods reported production increases; however, the number of employees decreased at 10 of these plants.

Methodology of the Study

INFORM's research on reduction of industrial hazardous and toxic wastes at source came primarily from interviews and correspondence with high-level plant officials at each of the study plants, but also from published materials of various kinds. These publications included corporate annual reports and other company literature, Resource Conservation and Recovery Act (RCRA) biennial report forms, Toxics Release Inventory report form R's, New Jersey Hazardous Waste Minimization Survey forms, and New Jersey Right-to-Know report forms.

In the second round of research, which began in 1987, INFORM gathered data on source reduction activities that took place during the entire period — predominantly from 1978 through 1990. That is, officials from plants that were providing information for the first time (because they had not granted INFORM interviews in the first phase of the research) had the opportunity to report on activities that took place prior to the *Cutting Chemical Wastes* study.

The Study Plants

For the original 1985 *Cutting Chemical Wastes* study, INFORM chose plants from three of the largest hazardous waste-generating states (California, Ohio, and New Jersey). The plants were selected to be broadly representative of the wide diversity of facilities across the organic chemical industry.

Thus, the plants vary in size (from Colloids, in California, with four employees to Du Pont Chambers Works, in New Jersey, with 3,500), in age (for example, the Merck plant in New Jersey was built in 1903, while the PMC plant in Ohio was built in 1966), in the types of products they produce, and in the type of process they use (batch or continuous). (Continuous processing dedicates a given set of equipment to the continuous production of a single product, while batch processing uses a given set of equipment to manufacture a variety of products at different times. Continuous processing generally generates less waste per pound of product because large volumes of the product can be produced without frequent start-ups, shut-downs, and cleaning of the equipment between batches.)

While these three states are similar in that each has a substantial chemical industrial base, they differ in regulatory climate. California hazardous waste laws tend to go further in terms of the number of chemical wastes regulated and waste disposal restrictions — the state added the so-called "California List" of materials (see Appendix D) to waste regulated in that state under the federal Resource Conservation and Recovery Act (RCRA), and imposed land disposal restrictions ahead of the US Environmental Protection Agency's timetable. California laws also tend to be implemented at the local government level to a greater extent than in the other two states.

New Jersey's environmental laws are known for their innovation. Examples include the Toxic Catastrophe Prevention Act (TCPA), the Environmental Cleanup Responsibility Act (ECRA), and the New Jersey Right-To-Know law. The 1980 New Jersey Industrial Survey served as the basis upon which the US Environmental Protection Agency's Toxics Release Inventory was structured.

Ohio's environmental laws were the least restrictive at the time of the INFORM study. For example, Ohio was one of the few states in the nation allowing deep well injection of hazardous waste.

Table I-1 provides basic information about all 29 study plants: location, size (number of employees), production level changes between the two phases of the study, reported source reduction, cooperation with INFORM, and products manufactured.

The plants revisited in the second round of research have been the subject of considerable public discussion as a result of the publication of *Cutting Chemical Wastes*. Thus, their managers may be, on average, more aware of source reduction and its potential than officials at other chemical plants around the country. The amount of source reduction occurring at these plants is, therefore, not necessarily representative of the industry as a whole.

Cooperation with INFORM

Thirteen of the 29 plants cooperated with INFORM for the first round of research resulting in *Cutting Chemical Wastes*. Of the original 29 plants, 27 are still in manufacturing: Frank Enterprises is now strictly a sales office, and the Fibrec plant no longer operates. In this second round of research, 19 plants fully cooperated in the study. Five additional plants partially cooperated, in that they provided some information through written correspondence or by telephone, but did not grant an on-site interview. This greater cooperation with INFORM during the second round of research has made possible a more thorough analysis of source reduction trends and impacts over a period of about a dozen years.

The Plant Interview

INFORM developed a standard questionnaire as the basis for interviews with plant managers. It sought information on both source reduction programs and source reduction activities and accomplishments, and covered seven basic topics.

1. **Responsibility for source reduction policy**
   - Does your company/plant have an official, written policy on toxic and hazardous waste management, and does it prioritize management options?
   - Where does source reduction stand in this policy?
   - Which division, office, or person oversees the implementation of this waste management policy? Of the source reduction component?

2. **Mechanisms to implement source reduction**
   - What are the elements of your plant's source reduction program? (See Box 2 for explanation of program features.)
   - Does the plant perform materials tracking to identify the sources, types, and amounts of hazardous and toxic wastes? What is its scope?
   - Is there a full cost accounting system for toxic and hazardous waste?
   - Who, within the plant, is responsible for reducing the generation of toxic and hazardous waste?

3. **Changes in plant activity**
   - What changes, if any, have there been in employee numbers, plant facility size, production levels, types of products, or ownership since the profile in *Cutting Chemical Wastes* was written?
   - How have these changes affected the overall generation of toxic and hazardous waste?

4. **Changes in source reduction activity**
   - What actions have been taken to reduce the hazardous and toxic waste generated by the plant since the profile in *Cutting Chemical Wastes* was written? For each source reduction action:
     - Why was the action taken?
     - What factors affected the decision?
     - How much did it cost to implement?

Box 2: Source Reduction Program Features

Throughout this report, the extent to which plants have adopted these source reduction program features is described as "full," "partial," or "none." These descriptions distinguish these categories.

Written source reduction policy A policy that promotes the primacy of source reduction as a waste management strategy in any explicit, written statement. A full source reduction policy is one that places source reduction as the top strategic waste management priority (ahead of recycling, treatment, and disposal) and identifies it as the preferred option in all cases. A partial source reduction policy puts source reduction on an equal footing with other waste management options.

Leadership Management-level responsibility for ensuring source reduction progress. Source reduction leadership can come from the plant manager or other nonenvironmental managers (such as a technical superintendent) or from an environmental or safety and health officer, but it is most effective when it comes from a combination of these.

Materials accounting A tracking of the inputs of individual chemicals (the amount existing in plant inventory plus the amounts entering the plant and created in plant processes) and the outputs (the amount consumed in plant processes plus the amount shipped out in, or as, primary product and the amount leaving processes as co-products or as waste). Materials accounting procedures are necessary for a full accounting of waste-related costs. Full materials accounting is multimedia (applies to all chemicals, whether solid, liquid, or gas) and chemical-specific. It identifies sources of wastes and activities leading to waste generation. Partial materials accounting includes some, but not all, of these procedures.

Materials balance Quantitative assessment of chemical inputs and outputs of individual processes to determine if all sources of waste have been identified. A materials balance is a more rigorous form of materials accounting. It aims to account for every pound of a chemical that is (a) shipped to the process, (b) created or destroyed in the process, (c) delivered as a product from the process, and (d) wasted (regardless of whether it is an air, water, or solid waste). If the amount of waste identified does not equal the difference between the amount of the chemical entering (or being created in) and leaving (or being consumed in) the process, other sources of waste probably exist. Such a materials balance is multimedia and chemical-specific and includes all inputs and outputs.

Full cost accounting Accounting procedures that incorporate pollution costs (such as the costs of pollution control, waste disposal, regulatory compliance, lost materials, insurance, future liabilities, and public and customer relations dealing with waste issues) into the plant's cost accounting system and assign the costs to individual processes, rather than to general plant overhead. Full cost accounting is multimedia and chemical-specific, and is done at the process level, while partial cost accounting includes some but not all of these criteria. (Plants identified in the findings as having full or partial cost accounting systems may not include all of the waste-related costs identified by INFORM in their system.)

Employee involvement Programs to involve employees at all levels (from top managerial staff to production and maintenance workers) in the company's source reduction activities. Management can encourage employee involvement in a variety of ways: by soliciting ideas from them, by offering them source reduction training, and by rewarding them for suggesting and/or implementing successful source reduction projects.

Environmental goals Specific goals set for reduction of the generation or release of specific chemicals and wastestreams (although not necessarily explicitly through source reduction since the INFORM study plants with goals did not generally identify the strategies to be used for achieving them). Goals can be set for reductions in generation of specific chemical wastes from specific processes or from the plant as a whole, and can be set for individual environmental media (air emissions, wastewater discharges, or hazardous solid wastes) or for all media.

Environmental program INFORM also looked at whether or not a plant had, in addition to these source reduction program features, a formal program dedicated to environmental issues in general. For the purpose of estimating the impact of such a formal program on source reduction activities at the plant, INFORM distinguished the extent to which source reduction played a role in these programs by describing an environmental program as "full" if it included specific mention of source reduction as an option.

Box 3: Source Reduction Techniques

Process changes involve refinements or alterations in the chemical reaction process itself. They can range from simple changes of process conditions, such as temperature or pressure, to use of new chemical pathways or production techniques that can advance the state of the art in manufacturing as they help to achieve source reduction.

Operations changes involve improving plant operations, including material handling and equipment maintenance, in order to create less waste. They can include better control of material use and employee practices in order to minimize spills, process upsets, the excessive use of chemicals, or other problems that can generate wastes. Operations changes can occur in every stage of the manufacturing process, including storing, moving, mixing, and reacting chemicals.

Equipment changes, modifications, and additions can also occur in every stage of the manufacturing process, including storing, moving, mixing, and reacting chemicals. Because equipment is used in all aspects of a plant's operation, there are numerous opportunities for waste generation and as many chances to implement source reduction through equipment changes.

Chemical substitutions involve using raw materials that create fewer toxic and hazardous wastes during the production process without necessarily changing the product itself. Furthermore, chemical manufacturing facilities use many materials for essential operations outside the manufacturing process itself, such as cleaning and maintenance, pollution control, and corrosion inhibition. Substituting nonhazardous or nontoxic chemicals for these purposes can also reduce the generation of hazardous and toxic wastes.

Product changes involve redesigning the end product so its manufacture creates less toxic and hazardous waste and can often be achieved without changing the fundamental manufacturing process. For example, creating a chemical product in the form of pellets rather than as a powder can reduce the generation of waste dusts as the material is packaged.

- Was money saved? If so, how much?
- What technique was used? (See Box 3 for explanation of source reduction techniques.)
- What year was it implemented?
- What was the name(s) of the specific chemical waste(s) reduced?
- How much was reduced (as pounds and as percentage)?
- What effect, if any, was there on product yield?
- How much time was needed for implementation (including research and development)?

5. **Information transfer**
   - Have you received technical assistance from anybody and have you offered technical assistance to others on the subject of source reduction? Why or why not?

6. **State and federal impact on source reduction**
   - Is your source reduction program being encouraged/hindered by federal or state statutory/regulatory requirements (e.g., RCRA waste minimization reporting, Clean Water Act or Clean Air Act requirements)?

7. **Future**
   - What are the future source reduction goals at your plant?

Following the interviews, a profile of each study plant was drafted containing information from all sources on the plant's source reduction practices. These drafts were sent to the plants to be reviewed for accuracy. While not all of the 24 fully and partially cooperating plants responded, those that did provided corrections and additional descriptions of source reduction practices or confirmation of the facts presented in the profiles. Since a year or more had passed between the interview date and the review period, each plant was requested to provide updated information on source reduction programs and activities. Only Exxon, of the plants that initially cooperated with INFORM's study, failed to respond to this final follow-up request.

Plant managers did not provide data for each category of interest for each of the 181 source reduction activities identified. (In many cases, companies do not maintain records on such information.) Therefore, a different subset of source reduction activities and plants is used in reporting each finding. Table I-2 shows the number of source reduction activities with data in each of the categories mentioned above.

Table I-2: Number of Source Reduction Activities (SRAs) for Which Plants Provided Information, by Category (total of 181 SRAs)

Information Category	Number of SRAs for Which Data Were Provided
Source reduction technique used	177
Motivating factor	162
Waste medium affected	155
Year implemented	143
Percent reduced	99
Amount reduced	80
Dollars saved annually	62
Dollars spent (capital costs)	48
Payback period	38
Implementation time (including research and development)	33
Effect on product yield	20

Scope of the Study

This study is designed to provide readers both with an analysis of chemical industry source reduction progress and with detailed information about specific source reduction activities at each of the study plants. To that end, it consists of two parts and a series of appendices.

Part I includes this introduction and a chapter on the overall findings and conclusions of this study. The findings are broken down into four basic groups: source reduction accomplishments at the study plants, source reduction program features and plant characteristics, motivation for source reduction activities, and source reduction techniques used. The analysis identifies overall source reduction impact, patterns among the study plants, and trends over time.

Part II presents individual profiles of each study plant. Each profile includes a summary and information about the plant's products and operations, environmental policy, materials data collection methods, source reduction program, source reduction activities, other waste management practices, and technical assistance, as well as plant officials' comments on state and federal regulatory requirements. In addition to this narrative material, each profile contains a table summarizing key information about each source reduction activity implemented at the plant that was reported to INFORM since the publication of *Cutting Chemical Wastes* (137 source reduction activities in all). The 44 source reduction activities that were described in that earlier report are included in the analysis in Part I, but are not discussed in the profiles in Part II. Readers may refer to the profiles in *Cutting Chemical Wastes* for detailed information about these activities.

Finally, the appendices include an explanation of the statistical tests used on the data obtained from the plants, a discussion of the methodology used to estimate the impact of source reduction activities on waste generation and on Toxics Release Inventory releases and transfers, a bibliography, and a glossary of technical terms used in this text.

PART I CHAPTER 2

Findings and Conclusions

INFORM's new research identified a total of 137 individual source reduction activities. These activities took place at 21 of the 29 plants covered in INFORM's earlier study. Two of the original plants (Fibrec and Frank Enterprises) have since closed, or are no longer in manufacturing, three plants did not cooperate with the new research (Bonneau Dye, Hart Chem/J. E. Halma, and Shell), and three did cooperate but reported no source reduction activities (Max Marx, Morton, and Unocal). The total of 181 source reduction activities analyzed in this chapter includes 44 practices that INFORM documented in *Cutting Chemical Wastes*, as well as the 137 practices documented in this second round of research.

Based on the information gathered about these activities, INFORM's findings fall into four main areas:

1. Source reduction accomplishments at the study plants
2. Source reduction program features/plant characteristics
3. Motivation
4. Source reduction techniques

In addition, the findings include an analysis of the effect of source reduction on Toxics Release Inventory (TRI) releases and transfers reported by the study plants.

Based on the findings at the study plants, INFORM has drawn a series of conclusions about how toxic and hazardous waste-generating facilities throughout the United States (and, indeed, the world) can dramatically decrease their generation of these wastes, while saving money and increasing production efficiency.

Table I-3 summarizes the data obtained for each of the 27 plants still in manufacturing: the number of source reduction activities at each plant, the average percentage of individual targeted wastestreams reduced, total amount of waste reduced, the total dollars saved annually, the total amount of capital invested, the average payback period, the average percent change in yield, and the average time needed for implementation (including research and development). The table also indicates the number of source reduction activities for which plants provided data in each of these categories, as well as the number of source reduction activities for which plants provided data about source reduction techniques, type of waste reduced, product yield changes, and motivating factors. The remainder of this chapter provides additional information about the data summarized here.

Note about the data: In the findings that follow, figures showing total number of source reduction activities per year include data from 1978 through 1988 only. This is because data for the years 1989 and 1990 are not as comprehensive as those for the previous years; the initial in-depth interviews were conducted in 1987 and 1988, with updates through telephone conversa-

Table I-3: Summary of Source Reduction Data by Plant*

Plant Name	Number of Source Reduction Activities	Average Reduction (%)	Total Amount Reduced (lb/yr)	Total Dollars Saved (per year)	Total Capital Investment	Average Payback Period (months)
Large Plants						
Aristech	16	64% (5)	26,929,000 (6)	$3,754,800 (6)	$9,492,200 (7)	16.0 (4)
Atlantic	9	33% (1)	350,000 (1)			
Chevron	5	66% (3)	140,000 (1)	$200,000 (1)		
Ciba-Geigy	16	76% (11)	293,000 (6)	$1,593,100 (8)	$290,000 (8)	7.2 (8)
Dow	5	98% (3)	12,160,000 (1)	$2,726,000 (2)	$250,000 (1)	1.3 (1)
Du Pont	13	62% (4)	39,290,000 (10)	$3,755,000 (5)	$11,000,000 (2)	43.6 (1)
Exxon	9	72% (9)	17,089,810 (6)	$3,412,305 (6)	$18,700,000 (3)	47.9 (3)
Fisher	21	43% (8)	629,670 (15)	$529,000 (5)	$79,000 (5)	8.8 (4)
ICI Americas	7	67% (3)	125,566 (4)	$266,085 (4)		
IFF	11	30% (10)				
Merck	5	50% (1)	12,963,000 (4)	$1,047,750 (2)	$1,000,000 (1)	12.0 (1)
Monsanto	13	90% (7)	17,329,900 (5)	$3,715,850 (6)	$60,000 (2)	1.2 (2)
Morton						
PMC	8	85% (6)	90,510 (3)	$260,000 (2)	$25,000 (1)	
Medium Plants						
American Cyanamid	6	94% (5)	805,600 (5)	$220,000 (3)	$800,000 (5)	1.5 (3)
Borden	13	95% (13)	293,070 (6)	$46,620 (6)	$39,000 (5)	11.9 (4)
Def-Tec	2	88% (1)	2,535 (1)	$27,000 (2)		
Rhône-Poulenc	7	53% (3)	248,010 (6)	$365,500 (6)	$4,260,000 (5)	11.5 (4)
Shell						
Unocal	1	100% (1)				
Small Plants						
Bonneau Dye						
Colloids	3					
Hart Chem/J. E. Halma						
ICI Resins	3					
Max Marx	0					
Perstorp	7	72% (4)		$40,000 (1)		
Scher	1	100% (1)				

Notes:
1. Blank spaces indicate no data provided by plants.
2. Plants not included: Fibrec (plant no longer exists) and Frank Enterprises (no longer in manufacturing).
* Numbers in parentheses indicate the number of source reduction activities (SRAs) for which data in that category were available.

tions and written correspondence after that. Data for the years prior to 1978 are not shown in these same figures because reported source reduction activities were few at the INFORM study plants prior to 1978 and their implementation was scattered over a 20-year period.

However, figures showing averaged information (such as average dollars saved per source reduction activity) do include source reduction activities implemented in 1989 and 1990 because this information is specific for each activity, rather than each year.

Lastly, in three figures showing cumulative source reduction activities through 1988, data from the years prior to 1978 are included but combined into one entry called "pre-1978." Use of these cumulative figures allows a better illustration of trends in source reduction activities over time.

Average Yield Increase (%)	Implementation Time (months)	Data Provided on Source reduction techniques	Environmental media	Product yield up/down	Motivating factors
		(16)	(15)	(5)/(0)	(16)
8.0% (1)		(8)	(9)	(3)/(0)	(9)
		(5)	(3)	(3)/(0)	(3)
12.4% (6)	6.7 (6)	(15)	(16)	(6)/(0)	(12)
	5 (2)	(5)	(5)	(1)/(0)	(5)
0.0% (1)	3 (1)	(13)	(13)	(6)/(0)	(13)
	60 (1)	(8)	(9)	(5)/(0)	(8)
28.0% (1)	18 (1)	(21)	(8)	(9)/(0)	(17)
		(7)	(6)	(1)/(0)	(3)
7.0% (1)	3.5 (2)	(11)	(6)	(2)/(0)	(11)
		(5)	(5)	(4)/(0)	(5)
8.2% (2)	5.5 (2)	(13)	(12)	(4)/(0)	(13)
5.0% (1)	12 (3)	(8)	(8)	(2)/(0)	(8)
0.0% (1)	7.1 (6)	(6)	(6)		(6)
0.0% (4)	0.5 (4)	(13)	(13)	(4)/(1)	(13)
		(2)	(2)	(1)/(0)	(1)
1.0% (2)	6.5 (4)	(7)	(7)	(4)/(0)	(7)
		(1)	(1)		(1)
		(3)	(3)	(2)/(0)	(1)
		(3)	(1)	(3)/(0)	(1)
		(6)	(7)	(3)/(0)	(6)
		(1)			(1)

Source Reduction Accomplishments at the Study Plants

Overall, the INFORM study plants showed source reduction progress over the past decade. Breaking down the source reduction activities reported by the year in which they were first implemented shows a rise from less than 5 in 1978 to a peak of over 30 in 1987. As Figure 1 shows, prior to 1985 the peak was 9 source reduction activities (in 1982 and 1983) and this total was met or exceeded in each of the years from 1985 to 1988.

Figure 1: Source Reduction Activities (SRAs) by Year First Implemented

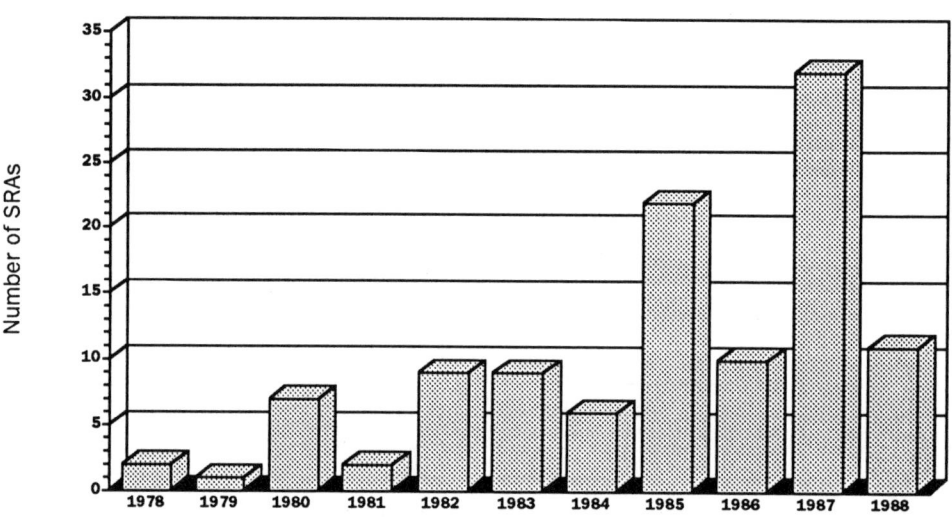

Year Source Reduction Activity First Implemented

Amount of waste reduced

More than one-third of the source reduction activities (28 out of 80) reduced between 10,000 and 99,999 pounds of waste, while another quarter (21 out of 80) reduced between 100,000 and 999,999 pounds of waste, and 18 percent (14 out of 80) reduced wastes by 1 million pounds or more (Figure 2). The total, for the 80 source reduction activities at 16 plants for which this information was reported, was 128.7 million pounds per year, or an average of 1.6 million pounds per source reduction activity.

Opportunities for reducing large amounts of waste through source reduction continue to be found. INFORM's study plants reported achieving large waste reductions per source reduction activity throughout the entire time period covered by the study, not just in the early years. Furthermore, not one plant official with an effective source reduction program indicated that the plant had reached its full source reduction potential, or even believed that a predefined level of source reduction can exist.

Percentage of wastestream reduced

More than one-quarter of the source reduction activities with percent reductions reported (29 out of 99 activities) achieved total elimination of the target wastestream, while more than half (51 out of 99) achieved reductions of 90 percent or more. The average, at the 20 plants reporting this information, was 71 percent reduction in the individual target wastestream. Figure 3 shows the number of source reduction activities reporting percentage reductions within given ranges.

An example of total elimination of a wastestream occurred at the Ciba-Geigy plant in Toms River, New Jersey, where the discharge of heavy metal wastes was eliminated by shifting the necessary purification steps to an earlier part of the process. Six dyes have been made using this method. It resulted in an increase in product output of 11 percent and an annual decrease in disposal of waste by 100 drums, at a cost savings of $86,500 per year. The action required no initial investment and was implemented in 2 months. (As described in its profile in Part II, the plant subsequently shut down its dye production and other manufacturing functions.)

Figure 2: Waste Reduced per Source Reduction Activity (SRA)

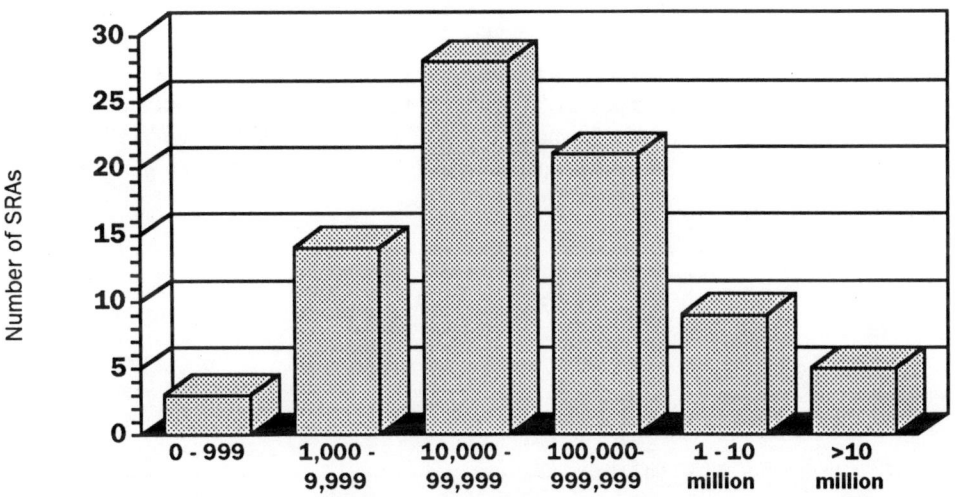

Waste Reduced per Source Reduction Activity (pounds)
(80 SRAs at 16 plants)

Figure 3: Percentage of Target Wastestream Reduced per Source Reduction Activity (SRA)

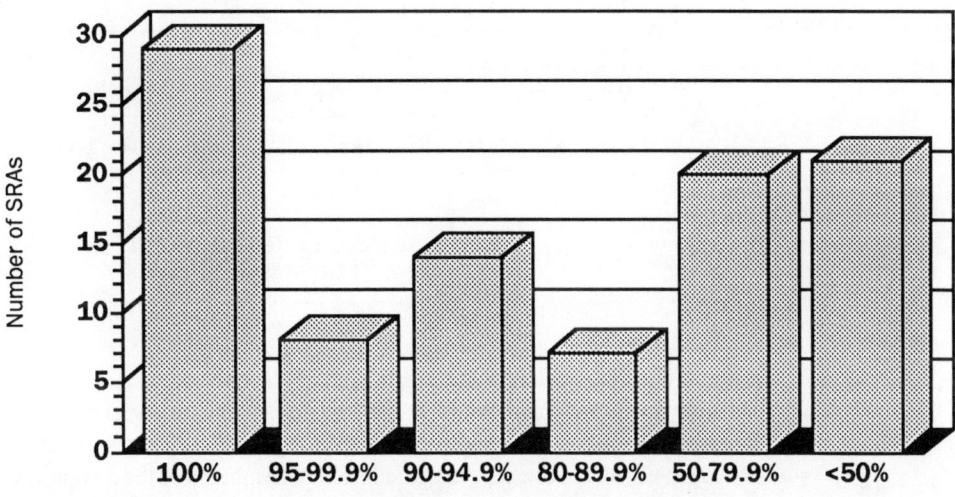

Reduction per Source Reduction Activity (percent)
(99 SRAs at 20 plants)

Medium of waste reduced

Nearly one-half (49 percent) of the source reduction activities with waste medium identified reduced wastewaters (75 out of 152), while 44 percent reduced solid wastes (67 out of 152), and 24 percent reduced air emissions (36 out of 152). (The totals come to more than 152 because some of the source reduction activities affected more than one waste medium.)

Figure 4 shows the cumulative number of source reduction activities per year occurring for each type of waste. Cumulative numbers show the number of source reduction activities in place in each year and thus better illustrate trends over time than the numbers of new activities alone. Further, the effects of source reduction activities are cumulative in the sense that wastes avoided in one year continue to be avoided in all subsequent years. (The numbers of source reduction activities shown for 1988 are less than the total numbers above because year of implementation was reported for only 114 of the 152 activities for which the medium reduced was identified.)

Source reduction activities affecting wastewater have predominated since the early 1980s. Solid wastes received little attention before 1981, perhaps because of widely available and inexpensive disposal options. But as the 1980s progressed, with increasing restrictions and liabilities associated with solid hazardous waste, the number of source reduction activities affecting them also increased.

Figure 4: Cumulative Number of Source Reduction Activities (SRAs) per Year, by Type of Waste

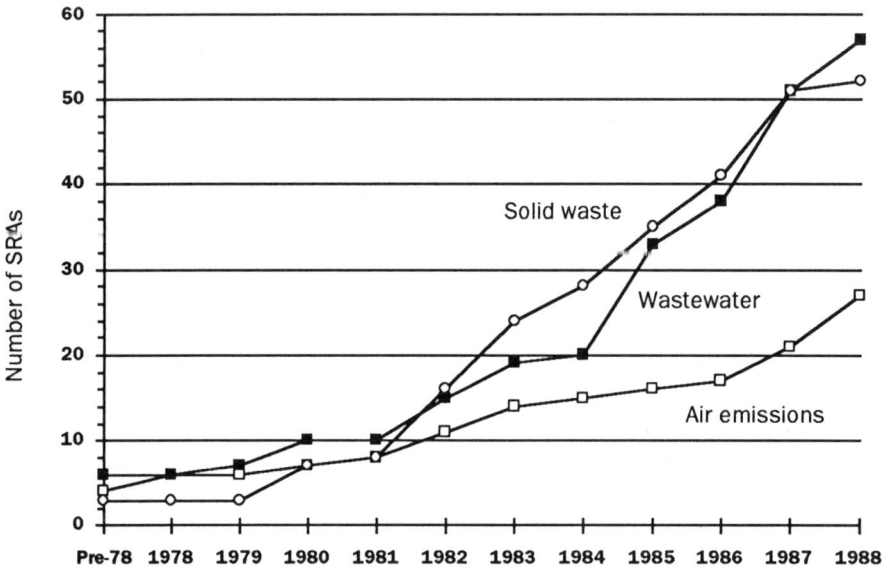

Year Source Reduction Took Place
(114 SRAs at 17 plants)

During this same time period, the trend in the number of source reduction activities for air emissions has not kept pace with the increases for wastewaters and solid wastes. This, however, is not due to lack of plants with air emissions. Of the 22 plants in this study with reports on toxic chemicals to the Environmental Protection Agency's Toxics Release Inventory (TRI) in 1987 and 1988, all but one reported air emissions. Eighteen of the 22 plants reported wastewaters and 16 reported generating solid wastes requiring disposal. While more plants reported TRI air emissions than releases to the other environmental media, fewer (14 out of 22) reported having undertaken source reduction activities affecting air emissions than measures affecting the other media. For wastewaters, 19 of the 22 plants reported efforts at source reduction, and for solid wastes 17 plants did. (It should be noted that air discharges of toxic and hazardous materials have been much less regulated than releases or transfers to other environmental media, and thus air disposal has often involved no regulation-related costs; the passage of the 1991 Clean Air Act should change this.)

Figure 5 shows the average percent reduction achieved per source reduction activity for each type of waste. On average, activities reducing air emissions reduced individual wastestreams the most (more than 80 percent), but the average percent reduction for each waste type exceeds 70 percent (the overall average, as discussed above, is 71 percent). Thus, the potential impact of source reduction appears to be independent of environmental medium.

Figure 5: Average Percent Reduction per Source Reduction Activity (SRA) by Type of Waste
(Numbers in boxes show number of source reduction activities reported for each type of waste.)

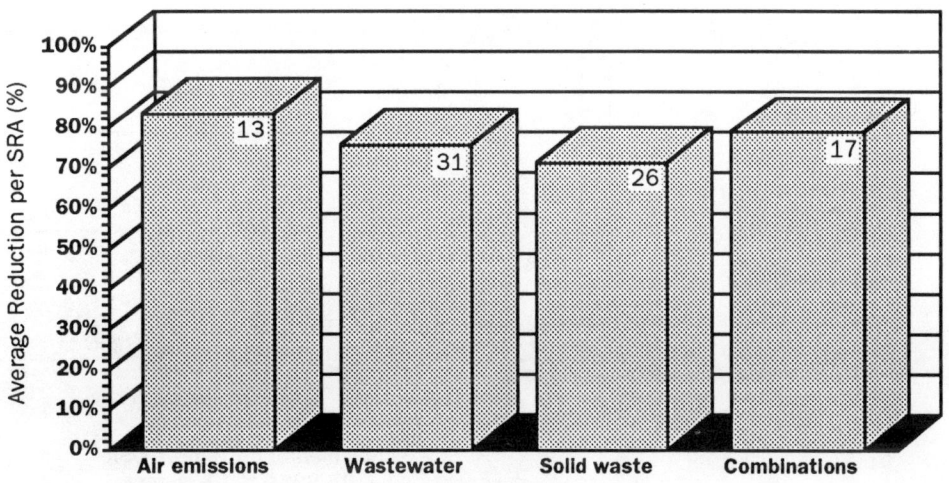

Types of Waste
(87 SRAs at 19 plants)

Implementation time

Nearly two-thirds of the 33 source reduction activities for which this information was reported (21, or 64 percent) required 6 months or less for implementation (including research and development); another 30 percent (10 out of 33) required 6 months to 3 years; and only 6 percent (2 out of 33) required more than 3 years for implementation (Figure 6). The average time to implement a source reduction activity was 8.2 months for 32 source reduction activities at these 11 plants (one series of source reduction activities with an overall implementation time of 20 years was not included in calculating the average implementation time since it lies so far beyond the time reported for the other 32 activities).

An example of a source reduction activity implemented over a short period of time occurred at the IFF plant in New Jersey. The corporate research and development group discovered alternate chemistry for producing one of its products. Through the use of a different catalyst system and different reaction conditions, it was found that the standard reaction could be run with better product yield, and that the resultant process wastestream would be free of the organic chlorides that were a by-product of the original process. Implementation of this change took about 3 months.

The series of source reduction activities with an overall implementation time of more than 20 years took place at the Monsanto plant in Ohio. In-process recycling of spent monomers was built into polymer production lines at the time of construction in 1972. Since then, process improvements have allowed an additional 1.4 million pounds of spent monomers from polymerization processes at the plant to be recycled and reused in the processes. These changes are a result of continuous research over the past 20 years to improve product yields.

Figure 6: Implementation Time per Source Reduction Activity (SRA)

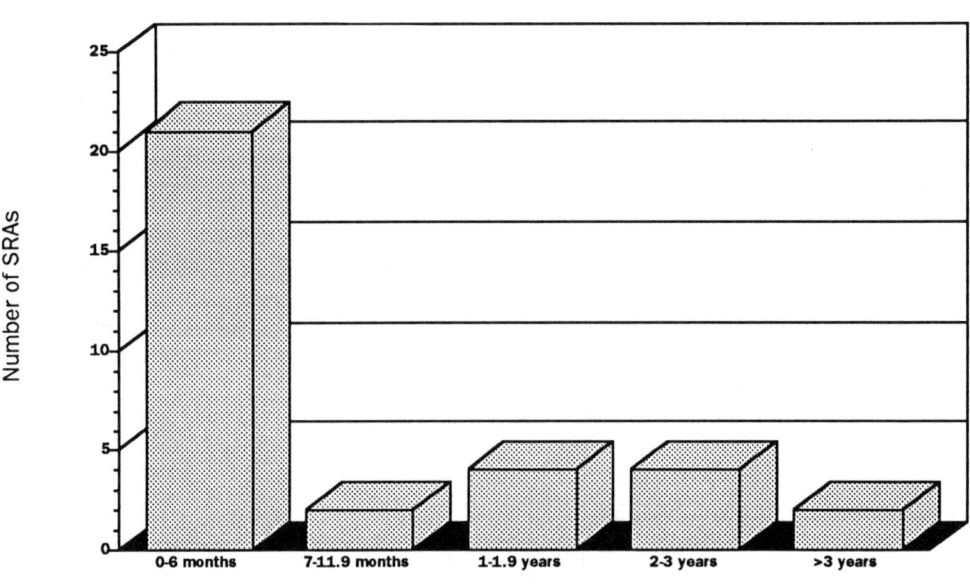

Implementation Time per SRA
(33 SRAs at 11 plants)

Changes in product yields

Sixty-eight of the 70 source reduction activities for which INFORM identified a change in product yield showed an increase in yield (one had no effect and one decreased the yield); of the 20 source reduction activities at 10 plants for which quantitative yield increase data were provided, 35 percent (7 out of 20) had yield increases between 10 and 40 percent, 25 percent (5 out of 20) had increases from 1 to 10 percent, and 40 percent (8 out of 20) had yield increases of 1 percent or less (Figure 7). The average increased production yield for these 20 source reduction activities was 7 percent.

The largest reported percentage increase in product yield was 40 percent at the Ciba-Geigy plant in New Jersey. Two changes in the multistep dye-making process made this increase possible and also eliminated wastes, resulting in an annual cost savings of $740,000. The first change, a chemical substitution, took place in the final step of the dye manufacturing process where a chemical conversion takes place. This conversion step was formerly carried out with iron as a raw material but, because of the large amount of solid iron sludge that formed, iron was replaced with a different conversion reagent. The second change was a process change in the conversion step. During the investigation of wastewater streams, the effluent from this process proved to be potentially toxic due to the presence of product in the wastewater discharge. Examination of the process led to improvements that eliminated this loss of product.

Figure 7: Percent Increase in Product Yield per Source Reduction Activity (SRA)

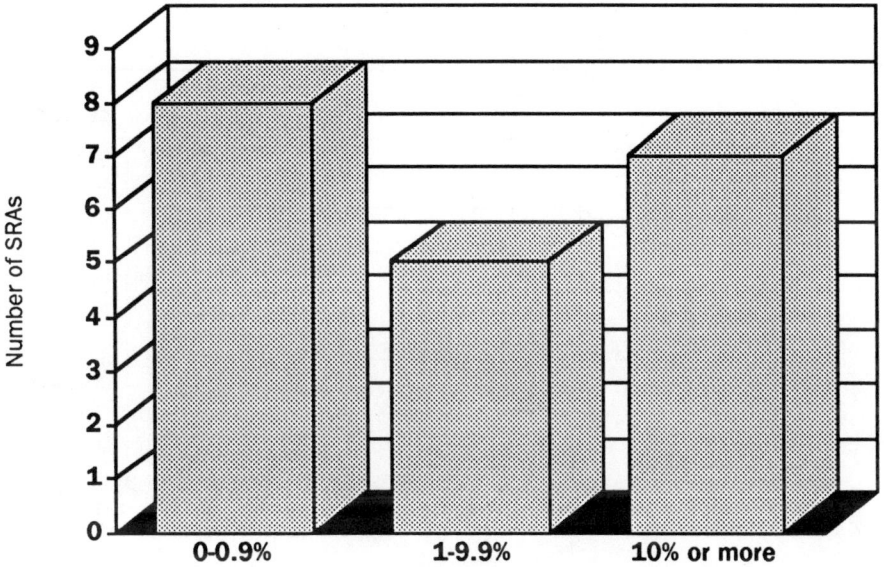

Percent Increase in Product Yield
(20 SRAs at 10 plants)

Savings

Fifteen percent of the source reduction activities for which annual dollar savings were reported (9 out of 62) save $1 million or more annually, nearly half (28 out of 62) save between $45,000 and $1 million annually, more than one-quarter (17 out of 62) save between $6,000 and $45,000 each year, and 13 percent (8 out of 62) save less than $6,000 annually; only one out of all 181 source reduction activities documented by INFORM reported a net cost increase due to a source reduction activity (Figure 8). The average annual savings per source reduction activity was just over $351,000 for these 62 source reduction activities at 14 plants for which cost savings information was reported. In all, these plants reported annual savings of $21.8 million.

The highest number of source reduction activities is in the $125,000 to $349,999 range, with most source reduction activities (34 of 62) falling between $16,000 and $349,999 saved per year. (The ranges were selected so that the high end of each category is roughly three times greater than the low end.)

A net cost increase was reported for only one source reduction activity. At the Aristech plant in Haverhill, Ohio, waste treatment costs increased by 40 to 50 percent when the company replaced chromium used for corrosion resistance in its cooling water with a newly available nonmetallic material.

Figure 8: Annual Savings per Source Reduction Activity (SRA)

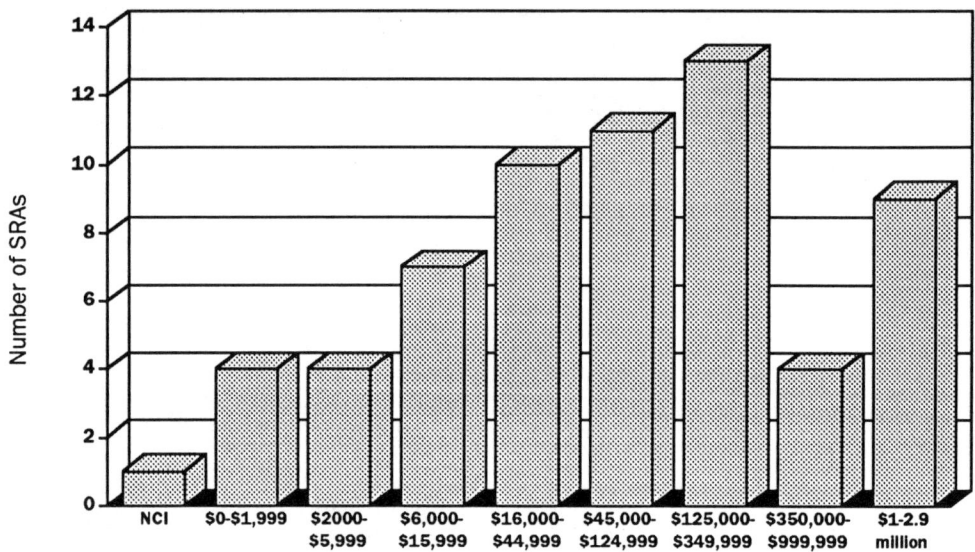

Annual Savings per SRA
(62 SRAs at 14 plants)
(NCI = net cost increase)

Capital costs of implementing source reduction activities

One-quarter of the source reduction activities for which capital cost information was provided (12 out of 48) required no capital investment for implementation; just under one-half (22 out of 48) required investments of less than $100,000, and these investments were recouped in savings in, on average, under 18 months (Figures 9 and 10). While 13 percent of the activities (6 out of 48) required investments of $1 million to $10 million to implement, these costs were paid back by savings in an average of 2.5 years. The source reduction project with the largest amount spent and the largest payback period in Figure 10 was undertaken by Exxon. It cost Exxon $18.7 million to replace filters with high-speed centrifuges to remove solids from lubricating oil additives. A second-stage separator was also added to recover the oil and active ingredients remaining in the centrifuge sludge. The annual savings of $1.56 million (a payback period of almost 12 years) represent recovered product and reduced disposal costs.

Figure 10 does not include the three highest payback periods reported to INFORM because placing these outlying data points on the graph would make all the other data points too small to see. These three are: 21 years (cost of $20,000), 70 years (cost of ($700,000), and 889 years (cost of $4 million) for source reduction activities implemented by Borden, Monsanto, and Rhône-Poulenc, respectively.

Figure 9: Capital Costs per Source Reduction Activity (SRA)

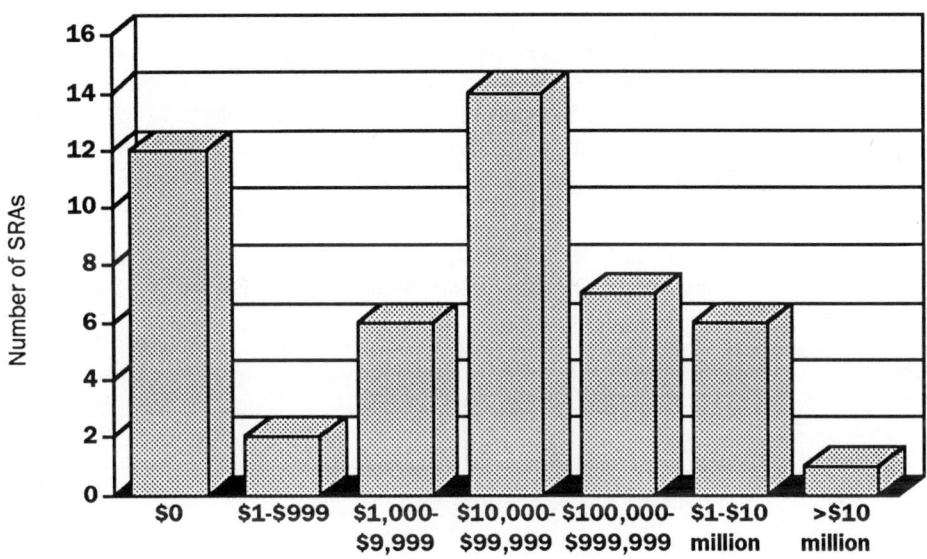

Capital Costs per Source Reduction Activity
(48 SRAs at 14 plants)

Figure 10: Average Payback Period per Source Reduction Activity (SRA) by Range of Capital Costs

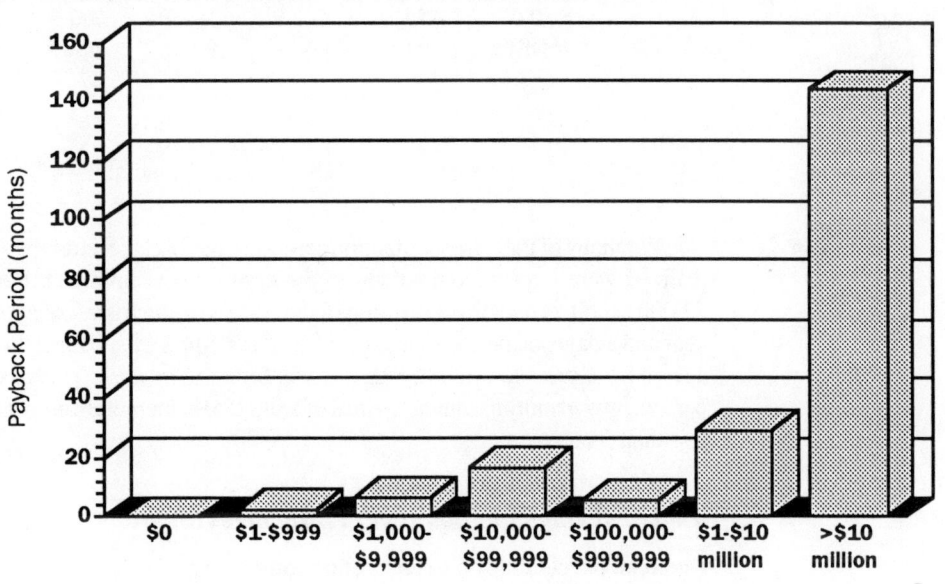

Capital Costs per Source Reduction Activity
(34 SRAs at 11 plants)

Payback period

Nearly two-thirds (24 out of 38, or 63 percent) of the source reduction activities for which payback period data were reported recouped their capital investments within 6 months or less; the payback period was 6 months to 3 years for another 18 percent (7 out of 38), 3 to 10 years for 8 percent (3 out of 38), and 10 years or more for 11 percent (4 out of 38) (Figure 11). The average payback period for investments made in source reduction activities was 13 months for 35 source reduction activities at 11 plants. (The three source reduction activities with payback periods of over 20 years were not included in this average because they fell considerably outside the payback range of the majority of the source reduction activities.)

Figure 11: Payback Periods per Source Reduction Activity (SRA)

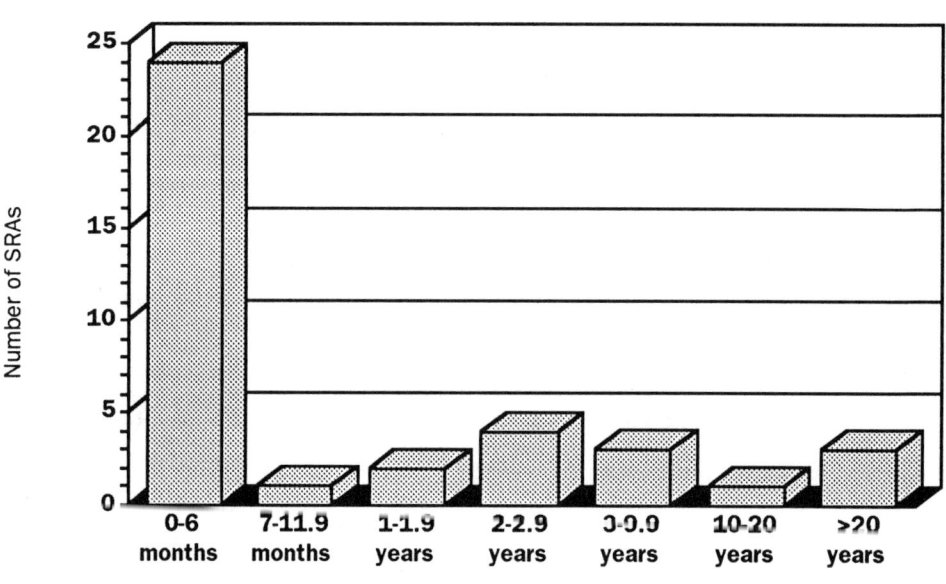

Payback Periods
(38 SRAs at 11 plants)

For many of the source reduction activities, payback periods were short because the savings realized were high in comparison to the capital costs involved. For example, Borden spent $2,000 to collect and distill flushings from truck loading filters for a savings of over $14,000 per year and a payback period of less than 2 months. Similarly, Monsanto spent $60,000 to upgrade a resin filter press in order to reduce leaks from the process. This has reduced the waste by 96 percent, with resulting annual savings of $300,000 in incineration costs. The payback period was less than 3 months.

Annual savings per dollar of capital investment

Nineteen percent of the source reduction of activities for which this information could be calculated (5 out of 27) reported annual savings of $8 or more per dollar of capital investment; another 37 percent (10 out of 27) reported annual savings per dollar invested of $1 to $6; and 44 percent (12 out of 27) reported annual savings of up to $1 per dollar of capital investment (Figure 12). The average annual savings per dollar spent on source reduction was $3.49 for 27 source reduction activities at eight plants.

One of the largest savings ratios was reported by Fisher Scientific, where a $500 investment to reuse solvents resulted in an annual savings of $5,250, or ten times the amount invested in one year. Dow achieved a similar return on a much larger scale, investing $250,000 in a process change to scrub a wastestream with water before using caustic, enabling the plant to reuse raw materials and thereby reduce waste, for annual savings of $2.4 million.

Figure 12: Annual Dollar Savings per Dollar of Capital Investment per Source Reduction Activity (SRA)

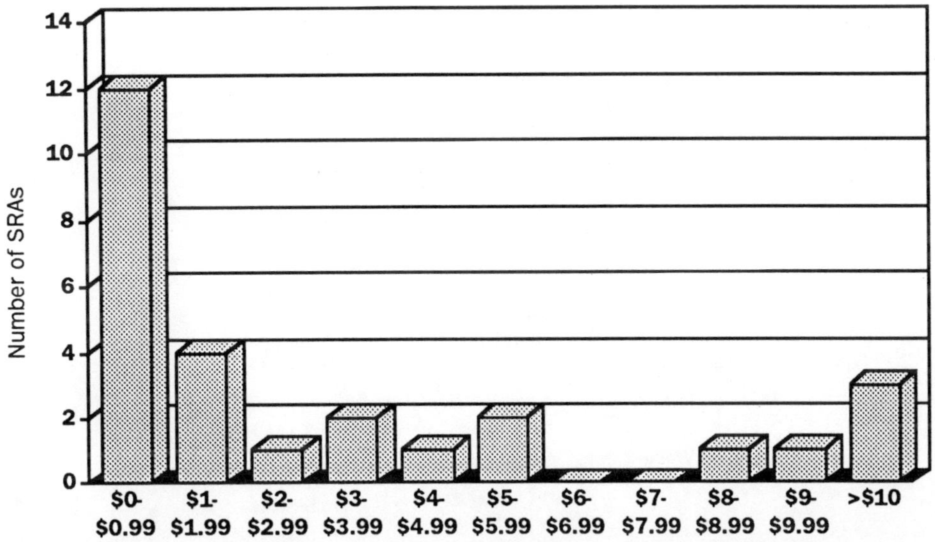

Range of Dollars Saved Annually per Dollar of Capital Investment
(27 SRAs at 8 plants)

INFORM found that, to date, the greater investments made in more costly source reduction projects were still producing real economic benefits and that the cumulative savings over time were extensive. Indeed, while five of the study plants (Aristech, Du Pont, Exxon, Merck, and Rhône-Poulenc) reported spending more than $1 million on source reduction projects to date, seven plants (Aristech, Ciba-Geigy, Dow, Du Pont, Exxon, Merck, and Monsanto) report net savings that now amount to over $1 million dollars annually (for plant-specific data, see Table I-3 on pages 12-13).

Source reduction activities in nonproduction functions

Source reduction opportunities exist in nonproduction as well as production functions of chemical plants. Nonproduction activities include raw material and product storage, loading and unloading procedures, cleaning and other maintenance operations, and pollution control equipment operations. Table I-4 summarizes information about source reduction activities, and their results, in production and nonproduction functions of the plants INFORM studied. Note that the data and percentages for different categories (such as pounds of waste reduced, or dollars saved) are based on different numbers of source reduction activities because the study plants did not provide information for every category.

Table I-4: Source Reduction Activities (SRAs) in Production and Nonproduction Functions*

	Production	Nonproduction
Number of SRAs	139 (81%)	33 (19%)
Number of plants	22	15
Total waste reduced (pounds)	127 million (98%) (69 SRAs)	2.2 million (2%) (11 SRAs)
Average percent reduced per SRA	70% (78 SRAs)	79% (16 SRAs)
Total annual savings ($)	$21 million (97%) (53 SRAs)	$545,000 (3%) (9 SRAs)
Capital costs ($)	$35 million (76%) (41 SRAs)	$11 million † (24%) (7 SRAs)
Average payback period (months)	13 months (32 SRAs)	8 months (4 SRAs)
Average implementation time (months)	9 months (28 SRAs)	5 months (5 SRAs)
Type of waste affected‡		
Air emissions	24 (21%)	8 (28%)
Wastewaters	59 (49%)	12 (41%)
Solid waste	47 (41%)	15 (52%)
	(115 SRAs)	(29 SRAs)
Technique used‡		
Operations changes	35 (26%)	21 (64%)
Equipment changes	22 (16%)	9 (27%)
Process changes	75 (55%)	3 (9%)
Chemical substitutions	16 (12%)	2 (6%)
Product changes	5 (4%)	0 (0%)
	(137 SRAs)	(33 SRAs)

* The total number of source reduction activities varies from category to category shown here because plants did not provide information about every category; percentages shown are based on the total number of source reduction activities within a category.

† One of the seven source reduction activities in this category cost $10 million; the other six combined cost $1 million. No payback period information is available for the source reduction activity costing $10 million.

‡ Numbers of SRAs may not equal totals because several techniques may have contributed to a single SRA, or a single SRA may affect more than one type of waste.

Source reduction accomplishments in nonproduction functions

Almost one-fifth of the source reduction activities for which INFORM could determine production or nonproduction function (33 out of 172, or 19 percent) took place in nonproduction functions of the plant, compared to 139 (81 percent) in production functions. As Table I-4 shows, source reduction activities in nonproduction functions accounted for 2 percent of the pounds of waste reduced (2.2 million pounds, for 11 out of 80 source reduction activities for which information on pounds reduced was available) and 3 percent of the annual cost savings ($545,430, for 9 out of 62 source reduction activities for which information on cost savings was available).

At the same time, source reduction activities in nonproduction functions reduced, on average, more of the targeted wastestream than did source reduction activities in production functions. They were implemented in half the time and showed a quicker return on investment, again on average. Thus, while there were fewer source reduction activities in nonproduction functions at the INFORM study plants, they offered, on average, fast results.

One activity in a nonproduction area used 22 percent of all the dollars invested in source reduction activities ($10 million), and six others used another 2 percent ($1 million in all), out of 48 source reduction activities for which capital investment information was available. The activity costing $10 million involved Du Pont's installation of a closed-pipe wastewater system (see Du Pont profile in Part II for a fuller discussion).

Payback periods/time needed for implementation

The payback period for recouping the initial investment for source reduction activities in nonproduction functions averaged 8 months (for 4 source reduction activities), and the time needed for implementation (including research and development) averaged 5 months (for 5 source reduction activities). For comparison, the payback period averaged 13 months for activities involving production functions (for 32 source reduction activities) and the implementation time averaged 9 months (for 28 source reduction activities).

Type of waste reduced in nonproduction functions

The type of waste affected by source reduction activities did not differ dramatically for production versus nonproduction functions; in both cases, air emissions were the least affected, involving about one-quarter of all source reduction activities. More specifically, source reduction activities in nonproduction functions affected air emissions 28 percent of the time (versus 21 percent for production functions), wastewaters 41 percent of the time (versus 49 percent for production areas), and solid waste 52 percent of the time (versus 41 percent for production functions). (These percentages are based on 29 source reduction activities in nonproduction functions and 115 in production functions; the percentages total more than 100 percent because some source reduction activities affected more than one type of waste.)

Source reduction techniques in nonproduction functions

Operations changes were involved in 64 percent of the source reduction activities in nonproduction functions, and equipment changes in 27 percent (see the section starting on page 48 on "Source Reduction Techniques" for a more detailed overall discussion of techniques). For comparison, process changes were the most frequently used technique for source reduction activities in the production process (used for 55 percent of all activities). (Note that percentages total more than 100 percent because several techniques may have contributed to a single source reduction activity.)

Trends in source reduction activities in nonproduction functions

Simple, low-technology source reduction activities in nonproduction functions continue to be found at the INFORM study plants. Figure 13 illustrates that, overall, the number of source reduction activities in production functions grew throughout the 1980s. The rise in the number of source reduction activities affecting nonproduction functions, while not as dramatic, indicates that when plant officials look for opportunities to reduce waste, they continue to find them, even in nonproduction functions.

Figure 13: Source Reduction Activities (SRAs) in Production and Nonproduction Functions
(Note: Scales are different for each graph)

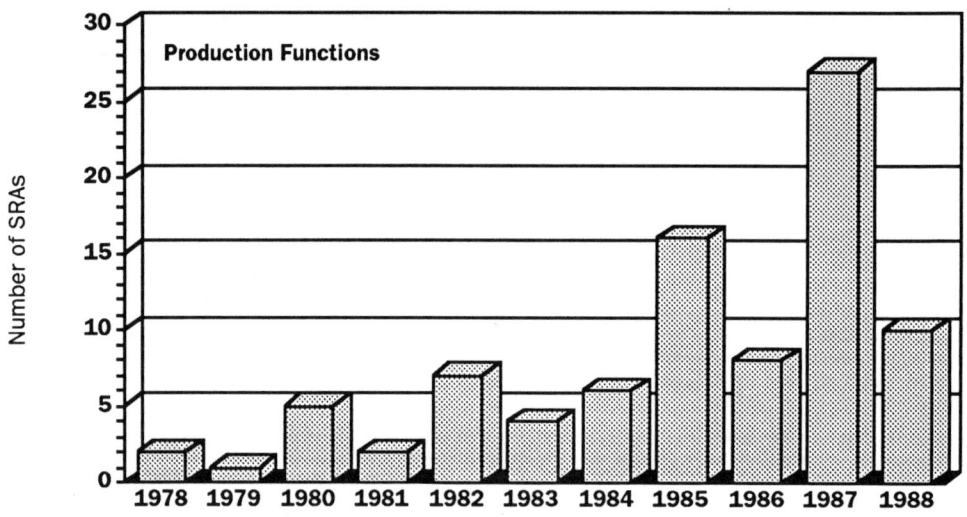

Year Source Reduction Activity Implemented
(139 SRAs at 22 plants)

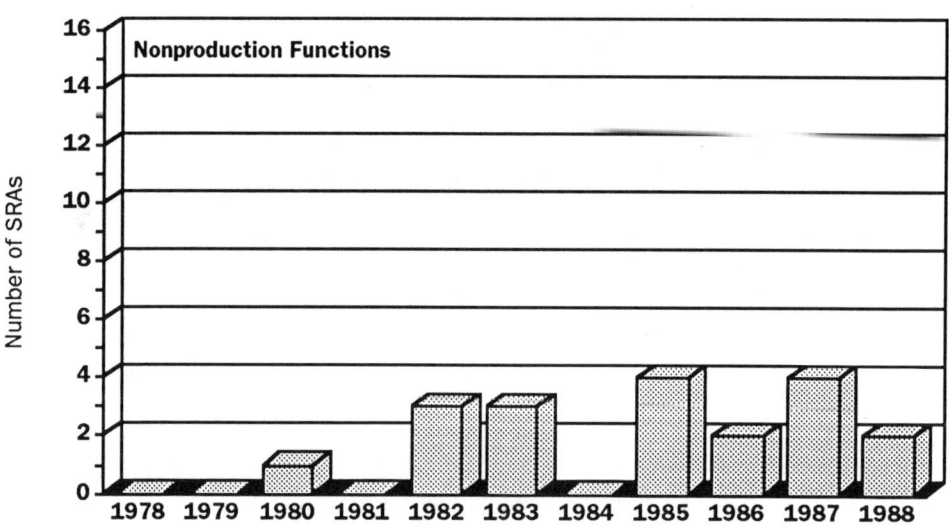

Year Source Reduction Activity Implemented
(33 SRAs at 15 plants)

Program Features and Plant Characteristics

Figure 1 (page 14) showed the source reduction progress made by INFORM's study plants through 1988. Particularly in the late 1980s, many of these plants implemented various components of a source reduction program. This report looks at eight specific features of such a program:

- written source reduction policy
- materials accounting
- materials balance
- cost accounting
- type of leadership
- employee involvement
- specific environmental goals, and
- whether or not an existing environmental program includes source reduction as an integral component.

Box 2 in Chapter 1 (page 8) provides a full discussion of each of these program features.

Table I-5 summarizes source reduction program information for the 27 INFORM study plants still involved in manufacturing. Information on all of the eight program features listed above was available for 20 of these 27 plants. Information on fewer program features was available for another five plants. No information was available from the remaining two plants. Blank spaces in the table indicate that plant officials did not provide information for a specific category. (See Table I-8 starting on page 36 at the end of this section for a more detailed chart of these program features at each plant.)

Table I-5: Source Reduction Program Features at INFORM Study Plants

Plant Name	Written Source Reduction Policy		Materials Accounting		Materials Balance	Cost Accounting	Leadership	Employee Involvement	Environmental Goals	Environmental Program		Number of Source Reduction Activities (SRAs)
Large Plants												
Aristech	N		Y	87	N	P	PM/E	T,R	N	Y	87	16
Atlantic	N		N		N	N	PM	N	N	N		9
Chevron	Y	85	Y	<72	Y	Y	E	I	W,S	Y	86	5
Ciba-Geigy	P		Y	86	Y	Y	PM/E	R	W	Y	84	16
Dow	Y	86	Y	84	Y	Y	E	I,T,R	W	Y	86	5
Du Pont	Y	80	P	82	N	Y	PM	I,T,R	A,W,S	Y	83	13
Exxon	N		P	82	N	P	PM	N	S	Y	86	9
Fisher	N		Y	86	Y	P	PM/E	T	N	Y	87	21
ICI Americas	P		Y			Y	PM	R,T,I	N	Y	87	7
IFF	N		P	87				R,I		Y		11
Merck	P		Y	86	Y	P	PM	T,I	A,W,S	Y	86	5
Monsanto	Y	82	P	82	N	P	PM	R	A,W,S	Y	82	13
Morton	P									Y	90	
PMC	N		P		N	P	E	R,I	N	Y		8
Medium Plants												
American Cyanamid	P		P	83	N	P	PM/E	T	A	Y	86	6
Borden	N		P		N	P	PM	R	W	Y	87	13
Def-Tec	N		P		N	N	PM	N	N	N		2
Rhône-Poulenc	Y	88	Y	79	N	Y	E	I	N	Y	89	7
Shell	P		P	86				N	N	N	84	
Unocal	N		Y			N	PM		N	N		1
Small Plants												
Bonneau												
Colloids	N		P		N	N	PM	N	N	N		3
Hart Chem/ J. E. Halma												
ICI Resins	Y	85	Y		Y	N	PM	T,R	S	Y	83	3
Max Marx	N		P		N	N	PM	N	N	N		0
Perstorp	N		N		N	P	PM	T	N	N		7
Scher	N		P		N	N	PM	N	N	Y		1

Key:
Blank space, information not provided by the plant.
Y, yes; **P**, partial; **N**, no; **number** (e.g., 85), year program feature implemented.
Leadership: **E**, environmental position; **PM**, plant manager or other nonenvironmental position.
Goals: **A**, set for air emissions; **W**, set for wastewater discharges; **S**, set for solid waste; **N**, no goals.
Employee involvement: **T**, employee training; **R**, rewards; **I**, ideas solicited from employees: **N**, no employee programs.

Definitions Used in Table I-5: Source Reduction Program Features at INFORM Study Plants

Written Source Reduction Policy

"**Yes**" means source reduction is clearly the top strategic priority of the plant's written policy. (If the policy places source reduction at the top of the hierarchy, but is not multimedia, it is still counted as "Yes.") "**Partial**" means source reduction is on equal footing with other waste management options. "**No**" indicates no written source reduction policy.

Materials Accounting

"**Yes**" refers to materials accounting that is multimedia and chemical-specific and that identifies sources of wastes and activities leading to waste generation. "**Partial**" refers to materials accounting procedures that do not include all of these criteria. "**No**" indicates there are no materials accounting procedures used.

Materials Balance

"**Yes**" indicates that the plant conducts a materials balance at each process as part of the plant's materials accounting procedures. The materials balance is multimedia, chemical-specific, and includes all inputs and outputs to a process. (Processes may include non-production area of the plant, or the non-production areas may be considered processes unto themselves for the purposes of materials accounting and materials balance.) Process inputs include the amount of a chemical brought to the process as a raw material, plus the amount created within the process. Output quantities include the amount of the chemical destroyed or converted to another chemical, and the amount removed from the process as or in a product, by-product, or waste material. Theoretically, the input quantities should exactly equal output quantities. The data used do not necessarily constitute the rigor of an engineering mass balance as defined by the National Academy of Sciences. "**No**" means that the plant does not conduct a materials balance at the process level. (A partial heading is not applicable here because anything less would be considered materials accounting.)

Cost Accounting

"**Yes**" indicates multimedia, chemical-specific cost accounting at the process level. "**Partial**" refers to cost accounting procedures that do not include all of these criteria. Neither "partial" nor "yes" necessarily includes all of the waste-related costs outlined by INFORM (see description of full cost accounting in Box 2 on page 8 in Chapter 1 for a full list of these costs). "**No**" indicates that no cost accounting is done at the facility.

Leadership

Leadership means that someone at the management level is responsible for source reduction progress. "**E**" indicates that source reduction responsibility lies with an environmental position such as an environmental officer. "**PM**" indicates this responsibility lies with a nonenvironmental position such as a plant manager or technical superintendent.

Employee Involvement

"**I**" means that management solicits ideas from employees. "**T**" indicates source reduction training is offered to employees, and "**R**" means that employees are rewarded for suggesting and/or implementing successful source reduction projects. "**N**" means no employee involvement exists.

Environmental Goals

Goals mean specific goals are set for reduction in the generation or reported release of specific chemicals or wastestreams (although source reduction is considered the preferred strategy by INFORM, these goals are not necessarily achieved through source reduction since the strategies for achieving goals were not generally explicitly specified as part of the goals themselves at the INFORM study plants). "**A**" refers to goals set for air emissions, "**W**" refers to goals set for wastewater discharges. "**S**" refers to goals set for hazardous solid waste generation. "**N**" indicates there are no environmental goals.

Environmental Program

"**Yes**" indicates that the facility has a formal environmental program in place that at least partially addresses source reduction; a year following it indicates the year the program was initiated. "**No**," followed by a year, indicates a program exists but does not include source reduction as an integral component. "**No**" without a year indicates that no program exists.

Impact of broadly adopting source reduction program features

While none of the 20 study plants for which complete information about program features was available fully adopted all eight program features that INFORM evaluated, the 14 plants that fully or partially implemented more than half of the program features reported, on average, 2.5 times as many source reduction activities (10 per plant) as the six plants that implemented fewer program features (3.7 per plant) (Figure 14). Figure 15 shows the number of plants that reported to INFORM that they have at least partially adopted each program feature. More than half the plants have adopted cost accounting, employee involvement, an environmental program, leadership, and materials accounting, while fewer than half have adopted materials balance, environmental goals, and a written source reduction policy.

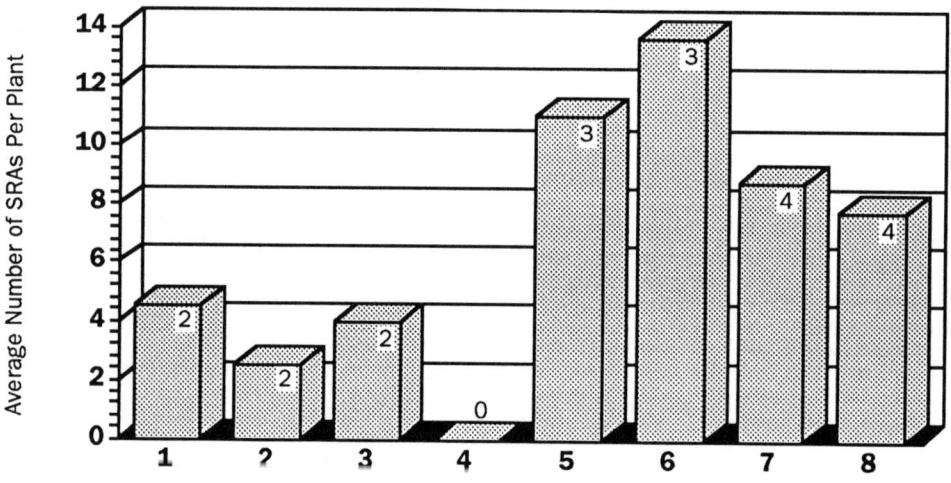

Figure 14: Average Number of Source Reduction Activities (SRAs) by Number of Source Reduction Program Features Established per Plant
(Numbers in boxes indicate number of plants with each number of program features.)

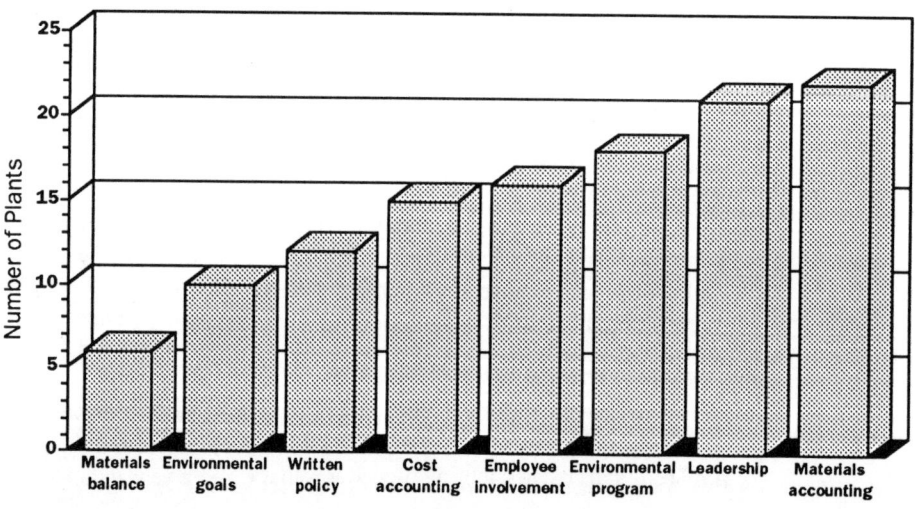

Figure 15: Number of Plants Fully or Partially Establishing Source Reduction Program Features

Table I-6: Average Number of Source Reduction Activities per Plant by Program Feature*

Program Feature	Average Number of Source Reduction Activities per Plant			Is the Difference between These Averages Statistically Significant? (yes or no)
	None	*Partial*	*Full*	
Cost accounting	2.7	10.9	8.8	Yes (between *none* and *partial* plus *full*)
Employee involvement	4.0	10.1	8.3	Yes (between *none* and *partial*)
	Plant manager/other	*Environmental*	*Environmental and plant manager/other*	
Leadership	6.1	6.3	14.8	Yes (between *plant manager/other* and *environmental and plant manager/other*) Yes (between *environmental* and *environmental and plant manager/other*)
	None	*Partial*	*Full*	
Environmental program	3.7	—	9.4	No
Source reduction policy	7.8	8.5	7.7	No
Materials accounting	8.0	7.2	8.6	No
Materials balance	7.6	—	8.0	No
Environmental goals	6.8	8.1	10.3	No

* *Based on information from 23 plants.*

Specific program features associated with greater numbers of source reduction activities

Plants that had one of three individual program features — cost accounting, employee involvement, and leadership from both environmental and other departments — had statistically significantly more source reduction activities, on average, than plants lacking these features. Table I-6 shows the average number of source reduction activities per plant by program feature, as well as the results of a statistical analysis of the correlation between the adoption of source reduction program features and the number of source reduction activities reported per plant. (For this analysis, the universe of plants included the 23 plants that provided information on any of the eight program features.)

To carry out the analysis, for each program feature, the number of plants adopting the feature and the number of source reduction activities at these plants were summed, and the average number of source reduction activities calculated. This average was tested for statistical significance using the Student's t-distribution to test for difference between means at the 95 percent confidence level. Correlation between two program features was tested using the chi-square distribution at the 95 percent confidence level.

A statistically significant difference means that, 95 percent of the time, a plant adopting this program feature will implement more source reduction activities than a plant that does not. The absence of a statistically significant difference means that the data available from the INFORM study plants were not sufficient to permit a statistically definitive conclusion (see Appendix A for further explanation of statistical significance and the specific methods used). For example, for the INFORM study plants, there is no statistically significant difference in the number of source reduction activities between plants with and without materials accounting. However, since only two of the study plants did not adopt even a partial form of materials accounting, it is difficult to make an accurate comparison.

Cost accounting

Plants with some type of cost accounting program (full or partial) had an average of three times as many source reduction activities (10.9 activities per plant for partial, 8.8 for full) as plants with no cost accounting system (2.7 activities per plant). Indeed, those plants with no cost accounting had the lowest average number of source reduction activities reported for any of the program feature categories.

Six plants have full cost accounting: Chevron, Ciba-Geigy, Dow, Du Pont, ICI Americas, and Rhône-Poulenc. Rhône-Poulenc's New Brunswick, New Jersey, plant, for example, started measuring waste per pound of product in 1979. It allocates all costs of waste treatment and disposal back to each process for all types of waste. Further, expenses for regulatory compliance, insurance, spill clean-up, and public/customer relations dealing with waste issues are assigned to each process as part of the full cost accounting at this plant.

Another nine plants do more limited cost accounting: American Cyanamid, Aristech, Borden, Exxon, Fisher, Merck, Monsanto, Perstorp, and PMC. Generally, such cost accounting includes disposal of RCRA wastes and/or wastewater treatment or sewerage charges, but not costs associated with air emissions or other regulatory and liability costs.

Employee involvement

Plants with some type of employee involvement had an average of more than twice as many source reduction activities (10.1 activities per plant for partial involvement) as plants with no employee involvement (4.0 activities per plant). Employee involvement programs include incentive reward programs, training, and ideas solicited from employees on a systematic basis. Three plants (Dow, Du Pont, and ICI Americas) have instituted all three types, while 12 other plants have one or two of these types of programs in place (American Cyanamid, Aristech, Borden, Chevron, Ciba-Geigy, Fisher, ICI Resins, Merck, Monsanto, Perstorp, PMC, and Rhône-Poulenc).

The corporate-wide source reduction program of Dow Chemical, for example, includes a recognition and reward system through which employees compete for awards. The winning projects are those with the largest cost savings, but the program takes a long-term view of costs and savings, recognizing "avoided costs" (such as liability or avoided manufacturing costs over 15 years) and not just one-time savings. In addition, Dow rotates production managers into its Environmental Quality Department as part of their training to ensure that future managers will be familiar with the problems and sensitive to the need for environmental quality.

Leadership

Plants with source reduction leadership both from the environmental department and from the plant manager or other nonenvironmental departments reported an average of more than twice as many source reduction activities (14.8 activities per plant) as plants with leadership from only the environmental (6.3 activities per plant) or only nonenvironmental (6.1 activities per plant) departments. Four plants (American Cyanamid, Aristech, Ciba-Geigy, and Fisher) had management personnel for their environmental programs from all aspects of plant operations.

Fisher Scientific in Fair Lawn, New Jersey, in particular, experimented with a committee to oversee its source reduction program that was limited to engineers and operational or research personnel. However, plant management found that without a full-spectrum, multidisciplinary committee, the program was not successful. Only when the committee was expanded to include representatives from the accounting department and from sales could the committee identify areas most costly in terms of waste and marketing opportunities for selling wastes as co-products.

In addition, all four of the plants with leadership from both environmental and other personnel also had cost accounting programs; these two program features may tend to reinforce each other in encouraging source reduction.

Other program features

While there were differences in the number of source reduction activities with respect to whether or not plants adopted any of the other five program features (written source reduction policy, materials accounting, materials balance, environmental goals, and an environmental program that includes source reduction), the differences are not statisti-

cally significant. However, this does not necessarily mean that these program features do not contribute to source reduction action at these plants as well.

Written source reduction policy and environmental program

While a written source reduction policy and an environmental program that includes source reduction are not by themselves statistically significantly associated with increased source reduction activity, the combination of these two basic features along with cost accounting and employee involvement did result in more source reduction activity per plant. As can be seen in Table I-5, those plants with some type of cost accounting generally also had environmental programs and source reduction policies. Further, as indicated earlier, plants that adopted more than half the source reduction program features had 2.5 times the number of source reduction activities as plants with fewer program features. Thus, while cost accounting is significantly correlated with implementing source reduction activities, at the INFORM study plants it is typically undertaken in conjunction with environmental programs and source reduction policies. Additionally, employee involvement programs appear at plants that also have environmental programs and cost accounting.

Materials accounting and materials balance

The available data do not make it possible to distinguish the effect of materials accounting or materials balance on the number of source reduction activities. Twenty-two of the plants indicated that they undertook some kind of materials accounting, and only two reported not doing any materials accounting. Further, since materials balance is a more rigorous form of materials accounting, all six of the study plants that reported undertaking materials balance also did materials accounting.

All but one (Perstorp) of the 15 plants with some form of cost accounting also had some form of materials accounting. Thus, while materials accounting by itself is not statistically significantly associated with a greater number of source reduction activities, it is often done at plants accounting for the full cost of waste generation. Intuitively, it would be difficult to do accurate and thorough cost accounting back to the source of waste generation without knowing where the source was in the first place.

Environmental goals

Only 10 of the 25 plants reporting program feature information have established environmental goals (specific targets for reduction of wastes or reported releases), and this feature has only recently been seen as a tool for implementing source reduction; thus, it may be too soon to identify what effect this program feature has on the number of source reduction activities. Those plants with environmental goals may have used them to identify source reduction opportunities but may not yet have had a chance to implement source reduction activities. The plants with environmental goals tend to also have a materials accounting system in place; this may aid in identifying and quantifying reduction goals.

Plant characteristics

Only size (based on number of employees), of three plant characteristics INFORM evaluated, was related to the number of source reduction activities at the study plant; the type of process used (batch or continuous) and changes in ownership did not make a difference in the average number of activities per plant. The first section of Table I-7 details these plant characteristics for each plant in the study, while the second part of the table presents the average number of source reduction activities per plant by plant characteristics.

Table I-7: Plant Characteristics and Number of Source Reduction Activities

Plant	Number of Source Reduction Activities	Type of Process	Year(s) of Change in Ownership
Large Plants (more than 100 employees)			
Aristech	16	Continuous	1986
Atlantic	9	Batch	1985/1989
Chevron	5	Continuous	None
Ciba-Geigy	16	Batch	None
Dow	5	Continuous	None
Du Pont	13	Batch/Continuous	None
Exxon	9	Batch	None
Fisher	21	Batch	1981/1986
ICI Americas	7	Batch/Continuous	1985/1986/1987
IFF	11	Batch/Continuous	None
Merck	5	Batch	None
Monsanto	13	Continuous	None
Morton	0	Batch/Continuous	None
PMC	8	Batch/Continuous	1985
Medium Plants (50-100 employees)			
American Cyanamid	6	Batch	None
Borden	13	Batch/Continuous	None
Def-Tec	2		1986/1988
Rhône-Poulenc	7	Batch	None
Shell	0		None
Unocal	1	Batch	None
Small Plants (fewer than 50 employees)			
Bonneau			
Colloids	3	Batch	1986
Hart Chem/J. E. Halma			
ICI Resins	3	Batch	1982/1985
Max Marx	0	Batch	1980/1988
Perstorp	7	Continuous	None
Scher	1	Batch	None
Total	181		

Average Number of Source Reduction Activities per Plant by Plant Characteristic*

Plant Characteristic	Average Number of Source Reduction Activities			Is the Difference between These Averages Statistically Significant? (yes or no)
	Batch	Continuous	Batch and Continuous	
Type of process	6.75	9.2	10.4	No
	Small	Medium	Large	
Plant size	2.8	7.0	10.6	Yes: between *small* and *large* No: between *small* and *medium* No: between *medium* and *large*

* Based on information from 23 plants.

Size

Large plants (more than 100 employees) reported an average of 10.6 source reduction activities per plant, while small plants (fewer than 50 employees) reported 2.8 per plant, a statistically significant difference. The difference between these groups and medium plants (50-100 employees), which reported 7.0 source reduction activities per plant, is not statistically significant.

Clearly, a larger plant having more processes, products, and/or wastestreams will have more potential source reduction activities than a small one. There is no way to know how many source reduction activities have been missed at each plant so that the total number of potential source reduction activities is not known. However, the adoption of each of the source reduction program features is independent of size. That is, a small or medium size plant was just as likely to adopt materials accounting or cost accounting, for example, as a large plant. Thus, the above analysis of the number of source reduction activities identified with each program feature should not be affected by plant size.

Process type

Differences in the average number of source reduction activities per plant for plants with batch processes (6.75 per plant), continuous processes (9.2 per plant), and both batch and continuous processes (10.4 per plant) are not statistically significant. While some discussion at plants with batch processing centered on the difficulty of reducing waste coming from ever-changing batch processes, this statistical analysis of INFORM study plants shows that the batch processing plants in this study, on average, implemented as many source reduction activities as those plants with continuous processes or with both batch and continuous processing (see Table I-7).

Change in ownership

There is no statistically significant difference in the average number of source reduction activities per plant between the nine plants that experienced changes in ownership during the period studied and those without ownership changes. However, several plants provided qualitative information on how the ownership changes affected source reduction progress.

For Aristech's Haverhill, Ohio, plant, divestiture from USS has enabled decisions to be made more quickly. The company also reports that, with the formation of a corporate environmental affairs department, instead of attorneys whose job was to ensure legal compliance, there is more knowledge about and support for source reduction efforts.

Def-Tec was also sold, by the Smith and Wesson Company. Def-Tec's manager explained that changes that have to be made are easier to accomplish without the many layers of command in a large company. On the other hand, the plant benefits from the many years of access to the resources and experience of a large company.

Several plants, Atlantic in Nutley, New Jersey, and Max Marx in Irvington, New Jersey, for example, were bought by other companies. They reported that very little has changed with respect to a source reduction program at their plants, in part because the companies that bought the plants did not make the same products as the plants, and so did not have source reduction expertise to contribute.

Table I-8: Source Reduction Program Features at the INFORM Study Plants

| Plant, State | Environmental Policy | | | Materials Data Collection | | Materials balance | Full cost accounting |
	Written policy with source reduction as top priority	Scope of policy	Date established	Materials accounting /Date established	Scope of materials accounting		
American Cyanamid, OH	Written policy to meet regulations. Includes source reduction, but not as distinct priority above treatment.	Corporate-wide	1977 with updates	Computerized database for materials tracking, cost: $500,000 /1983-86.	Manifested RCRA waste; tracks waste by process and by chemical. Capability to add constituents and air and water.	No	Cost accounting by product for RCRA and wastewater (air not allocated to process).
Aristech, OH	No written policy.			Statistical process control (SPC) /1984	All products	No	Cost accounting by process; for wastewater and solid waste.
Atlantic, NJ	No written policy.			No		No	No
Borden, CA	No written policy. Environmental manual including source reduction and energy conservation.	Corporate-wide	1989	Track product yields and do daily wastewater tank inventories.	Solid waste and wastewater	No	Cost accounting by product line, and by waste category for all environmental media (not chemical specific).
Chevron, CA	Written policy to "minimize" waste. Follow hierarchy with source reduction at top.	Corporate-wide	1985	Statistical process controls track pounds of raw material per pound of product.	All processes and products	Yes	Cost accounting by process for all media.
Ciba-Geigy, NJ	Written policy to meet regulations. Follow hierarchy with source reduction at top, but not in formal policy.	Corporate-wide		Materials tracking used to develop source reduction targets /1986 for wastewater.	All waste tracked by product	Yes	Cost accounting by process. Costs include waste disposal, compliance, insurance, cleanup, public and customer relations.

A blank indicates that the plant did not provide information.

Environmental Program				Employee Involvement		
Leadership	Program type	Date program established	Goals	Interdisciplinary committee	Progress reports	Training/ rewards/ideas
Corporate vice-president; plant environmental services manager	On-going program review, source reduction included in product development (hazard reviews), "waste minimization committee" in 1984.	1979, 1986	None plant-wide; targets for each project. The project at time of INFORM study was to eliminate largest air release.	"Waste minimization committee" of 4 employees with chemistry, public health and operations background.	Waste review of existing products every 3 years. Progress tracked against project targets.	
Corporate vice-president; plant's environmental services department	"Aristech Total Performance" (ATP) (quality program), employee suggestion program	1987, with change in suggestion program pre-1987	None	Waste minimization study and environmental audit teams use outside consultants and employees of other plants.	Progress on priority waste minimization projects sent to headquarters monthly.	Training in both attitudes and methods, employee suggestion program, awards for successful suggestions
Plant vice-presidents and plant engineer	On-going process improvements to increase yields. Substitute for unsafe chemicals with PMN approvals or eliminate use.		Eliminate use of chemicals that could cause health or environmental problems for workers and/or community.			None
Plant manager	"Waste reduction teams" for particular problems	1987	Specific plant goals; zero discharge of wastewater			Bonus for hourly employees tied to environmental goals
Environmental health and safety manager	Interdisciplinary teams review process operations weekly. Corporate program requires cost reduction goals (Save Money and Reduce Toxics, or SMART).	1985	Plant-wide in response to corporate goals; water and solid reduction by 50-60% corporate-wide goal by 1992.	Interdisciplinary teams of operations and maintenance departments, including environmental design and process engineers, to investigate process operations.	SMART program requires plant to report progress against reduction goals to headquarters.	All employees trained to use SPCs. Employees encouraged to bring ideas for product yield improvement to interdisciplinary teams
Plant manager, corporate office of environmental protection and services	Quality improvement process (QIP) includes source reduction, employee training, and motivation.	1984	For specific projects; in Toms River, reduction goals and strategies for wastewater have been the focus.	Task forces include chemists, engineers, and operations and maintenance personnel.	Progress report reviews bi-monthly (corporate staff with plant staff). Progress tracked against project targets.	QIP awards

(continued)

Table I-8: Source Reduction Program Features at the INFORM Study Plants (continued)

Plant, State	Environmental Policy			Materials Data Collection			
	Written policy with source reduction as top priority	Scope of policy	Date established	Materials accounting /Date established	Scope of materials accounting	Materials balance	Full cost accounting
Colloids, CA	No written policy			Annual audits and report on material losses		No	No
Def-Tec, OH	No written policy	Plant follows policy: "make no waste because it costs money."		"Scrap reports" list waste generated per unit of product./ Under Smith & Wesson ownership.	Solid waste	No	No
Dow, CA	Source reduction as top priority. WRAP (Waste Reduction Always Pays) program.	Corporate-wide	1986	Tracks pounds of each chemical per pound of product. /1984	Wastewater and solid waste, air added 1988-89	Yes	Costs of waste treatment/disposal tied to process/product. Compliance and liability insurance costs tied to facility overhead
Du Pont, NJ	Policy with source reduction first in hierarchy	Corporate-wide	1980	Tracks wastestreams by individual process, giving waste per 100 pounds of product. Subsequent measurements are at building units. Took $5 million and 20 person-years to establish. /1982	Wastewaters, solid waste, and recycled materials	No	Costs assigned to each process include capital costs, cost of lost raw material, public relations, compliance, treatment/disposal (based on waste standards), and environmental staff.
Exxon, NJ	No written policy			Annual corporate survey of solid wastestreams. Not clearly at source. One-time "waste minimization and and compliance" review./1982 for solid waste.	Annual corporate survey of solid waste per pound of product. 1987 "waste minimization review" for all media.	No; plant reports that because volumes are large, small errors result in large numbers	Costs of solid waste and wastewater tied to production unit; air emissions considered raw material losses. Other costs are compliance, insurance, clean-up, public/customer relations.
Fisher, NJ	No written policy			Computer-based materials handling system. Stress-strain monitors on storage tanks. /1983-1986.	Tank monitors for air. Computer system has component for all waste.	Yes	Costs of waste disposal are allocated back to originating process.

A blank indicates that the plant did not provide information.

Environmental Program				Employee Involvement		
Leadership	Program type	Date program established	Goals	Interdisciplinary committee	Progress reports	Training/ rewards/ideas
Plant manager	Good house-keeping practices		None			None
Plant manager	Cost reduction		None		No	None
Plant's environmental manager reports to major manager of production; corporate environmental quality department.	Corporate WRAP with guidelines and training for production managers. Recognizes avoided costs in approving projects.	1986 formalized	Yes, but not clear if project-specific, or overall plant or division, or corporate; goal of zero wastewater discharge.	Annual review of source reduction and recycling in combination for each plant at corporate level. Source reduction statistics include recycling.		Train managers at corporate level. Recognition and rewards for best projects.
Plant manager	"Waste minimization task force" for recommendations, quality achievement program, corporate capital funds for source reduction.	1983	35% reduction by 1990. Another 35% by 2000 for water/solid waste. 50% by 1993 for air. 90% by 1990 for carcinogens.	"Waste minimization committee," made up of environmental coordinators of all business units, makes recommendations and reviews progress.	Monthly within plant, annual to corporation	Monthly meetings on source reduction for all employees. New process approval review includes source reduction plans. Monetary awards for quality improvement.
Plant management-level person in charge of source reduction	Waste management program for each separate medium; emphasis on reducing land disposal.	1982-83	Reduce untreated waste going to landfills.	Waste minimization committee (1986) to look at joint refinery/chemical operations	Annual survey of waste to corporate headquarters	
Manager of environmental affairs and safety engineer	"Waste minimization committee" with multidisciplinary team. Waste management division.	1987	No	Multidisciplinary team including operations, research, financial, and sales staff	Generate database of source reduction achievments	Operator training to reduce solvent losses and cross-contamination

(continued)

Table I-8: Source Reduction Program Features at the INFORM Study Plants (continued)

Plant, State	Environmental Policy			Materials Data Collection			
	Written policy with source reduction as top priority	Scope of policy	Date established	Materials accounting /Date established	Scope of materials accounting	Materials balance	Full cost accounting
ICI Americas, CA	Written policy on waste handling and waste management. Source reduction not stated as top priority.	Corporate-wide		Waste tracked by chemical constituent	All media		Costs include waste-related costs by process; some indirect costs allocated.
ICI Resins, CA	Policy covers use of hazardous substances as well as spills, leaks, etc. Source reduction includes maintenance, engineering controls.	Plant-wide	1985	Measure input of raw materials and output of product when shipped. Weekly inspection of waste generated.	All batches	Easiest to measure inputs, yield is harder. Average over a year. Does not include waste.	No
IFF, NJ	No written policy			Establishing database, 1987	Solid and wastewater		
Max Marx, NJ	Policy to meet regulations	Plant-wide		Spot test of synthesis operations		No	No
Merck, NJ	Policy to meet regulations	Corporate-wide		For new products and processes /1986	All chemical inputs and outputs	For new products and processes	Includes treatment and disposal costs for liquid and solid waste associated with new products and processes
Monsanto, OH	For hazardous solid waste, hierarchy with source reduction at top. For air emissions, goal of zero discharges.	Corporate-wide	1970s, environmental guidelines; 1982, hierarchy	Solid waste inventories updated annually. SPCs in 1986.	Solid waste	No	Cost accounting for hazardous solid waste disposal costs. On-site treatment allocated to departments.

A blank indicates that the plant did not provide information.

Environmental Program				Employee Involvement		
Leadership	Program type	Date program established	Goals	Interdisciplinary committee	Progress reports	Training/ rewards/ideas
Technical superintendent and plant chemist	"Waste minimization program" for all wastes; committee to identify source reduction opportunities	1987	No	Committee has production, maintenance, environmental, and engineering personnel.		Monetary employee incentive program. Training program.
Plant manager and technical manager	Waste management handbook and surprise inspections with posted scores. Quality control training and re-education to change habits.	1983	Goals and timetables set for particular waste; this plant only generates solid waste.		No	Surprise inspections, quality control training based on "Quality Is Free" by Phil Cosby. Employees are scored based on inspection.
Corporate department of Environmental Compliance	"The Better Way" "Suggestion Award Program"					Financial awards for employee suggestions. Technical studies.
Plant operations manager	Housekeeping		None			None
Facility managers responsible for reducing amount of waste produced	Computerized simulation program, "PROVAL," for new batch processes.	1986	By end of 1991, reduce carcinogenic (and suspected carcinogens) air emissions by 90%. By 1993, eliminate them. By end of 1995, eliminate all toxic releases.			Corporate guidelines call for inventory control, "appropriate" use and disposal of chemicals, transmitting source reduction opportunities to environmental staff. (All salaried employees have "waste minimization" included in their performance evaluation starting in 1990.)
Plant manager, "waste reduction coordinator" monitor progress. (WRC is also senior environmental specialist.)	"Hazardous and solid waste minimization plan," "cost reduction program," "quality control program," "loss prevention and environmental control" review for new processes	1982	Annual reduction goals since 1982, 70% of all waste by 1992. Goal of zero air emissions as reported in TRI.		Progress reports against targets. Annual report on solid waste.	Annual contest for best quality control measures

(continued)

Table I-8: Source Reduction Program Features at the INFORM Study Plants (continued)

Plant, State	Environmental Policy			Materials Data Collection		Materials balance	Full cost accounting
	Written policy with source reduction as top priority	Scope of policy	Date established	Materials accounting /Date established	Scope of materials accounting		
Morton, OH	Adopted CMA "Responsible Care" guidelines	Corporate-wide	1990				
Perstorp, OH	Policy of optimal environmental protection, meet regulations by broad margins	Corporate-wide	1988	No		No	Sewer charges allocated to each process
PMC, OH	No formal policy			Weekly inventory of raw materials. Survey waste by product. /1981	Water	No	Sewer costs for wastewater charged to process, on historical basis; updated as processes change
Rhône-Poulenc, NJ	Policy with standards for environmental program, including source reduction	Corporate-wide	1988 "waste minimization standard"	Each process monitored to measure product yield, track raw material consumption and waste per unit product /1979	All waste	No	All costs allocated to process, including treatment, compliance, insurance, clean-up, public and customer relations
Scher, NJ	No formal policy			Inventory tracking of raw materials and products		No	No
Shell, CA	Written policy, including goals for source reduction	Corporate-wide		Source reduction and recycling inventories /1986	RCRA waste		
Unocal, CA	Policy covers compliance with regulations, no specific mention of source reduction.	Corporate-wide		Corporate computer-based materials tracking system			Environmental costs budgeted facility-wide, extrapolated from the previous year

A blank indicates that the plant did not provide information.

Environmental Program				Employee Involvement		
Leadership	Program type	Date program established	Goals	Interdisciplinary committee	Progress reports	Training/ rewards/ideas
	CMA's "Responsible Care Program"	1990				
Plant manager, corporate staff as consultants	Employee training		No			Annual OSHA training in handling of hazardous chemicals
Plant's environmental group	Cost reduction. List of chemicals not to use. Sewage treatment plant restrictions. Merit awards program.		No		No	Cooperation of chemists and environmental engineers through environmental group meetings for product development. Monetary awards for good ideas.
Plant's health and safety and environmental assessment supervisor. Corporate standards.	Annual corporate environmental report. R&D for new products looks at health and environmental effects of wastestreams.	Annual reports in 1989		Committees look at specific problems and include operations, technical, and environmental personnel.	Environmental review by corporate staff every 4-5 years; other periodic reviews	
Plant manager	Reuse or recycle any by-products or off-quality batches. Eliminate products using hazardous raw materials.		No			None
	Quality improvement program	1984				No
Plant manager and manager of environmental affairs	Review team for compliance with regulations		No		No	

Motivation

Factors motivating source reduction action

The problems and costs associated with waste disposal were the most frequently cited reason for implementing source reduction activities (cited for 67 out of 162 source reduction activities for which motivation information was provided, or 41 percent), followed closely by regulations limiting waste discharge or disposal options (cited for 58 activities, or 36 percent). As Table I-9 shows, less frequently cited reasons were product output (43 activities, or 27 percent), liability (32 activities, or 20 percent), and other (26 activities, or 16 percent). (The total number of source reduction activities exceeds 162, and the total of the percentages exceeds 100 percent, because multiple reasons were given for certain source reduction activities.)

The study plants provided INFORM with data on factors motivating source reduction for 162 of the 181 source reduction activities documented in this report. The categories of factors motivating source reduction action include: waste disposal costs, environmental regulations (such as more strict limits on quantities of wastewater discharge accepted by publicly owned treatment works, or banning of certain solid wastes from landfills), product output (such as production costs, improved product yield, customer requirements for certain product specifications, or product packaging), liability, and other factors not otherwise mentioned.

Table I-9: Factors Motivating Source Reduction Activities

Motivating Factor	Total Number of Source Reduction Activities*	Change over Time		Percent change
		Number of source reduction activities*		
		Pre-1985	1985-1990	
Waste disposal costs	67	16	35	+119%
Environmental regulations	58	10	40	+300%
Product output	43	9	19	+111%
Liability	32	12	18	+50%
Other	26	6	7	+17%

*Data were provided on 162 source reduction activities at 22 plants for motivating factors and on 119 source reduction activities at 19 plants for both motivating factor and year implemented; since years were not provided for all the source reduction activities, the total number of source reduction activities in the first column is greater than the sum of the source reduction activities in the pre-1985 and 1985-1990 columns.

Note: The total number of source reduction activities listed above exceeds 162 because multiple reasons were given for certain source reduction activities.

Trends in motivating factors

Although waste disposal has consistently been the most frequently cited factor motivating source reduction activities throughout the decade (increasing from the incentive for 16 source reduction activities implemented before 1985 to the incentive for 35 source reduction activities implemented between 1985 and 1990), other factors have grown faster at different points during the decade (Figure 16). (Note that the number of source reduction activities discussed here differs from the number discussed above because 22 plants provided information on motivating factors for 162 source reduction activities, while 19 plants provided information for 119 source reduction activities on both motivating factors and year of implementation.)

In Figure 16, the number of source reduction activities shown for each year is the total number of source reduction activities in place in that year, those implemented both during that year and in all previous years. It shows that the number of source reduction activities before 1980 was small, and the primary motivating factors were "other" reasons (such as employee suggestions) and waste disposal costs.

Figure 16: Motivation for Source Reduction Activities (SRAs) by Year (cumulative)

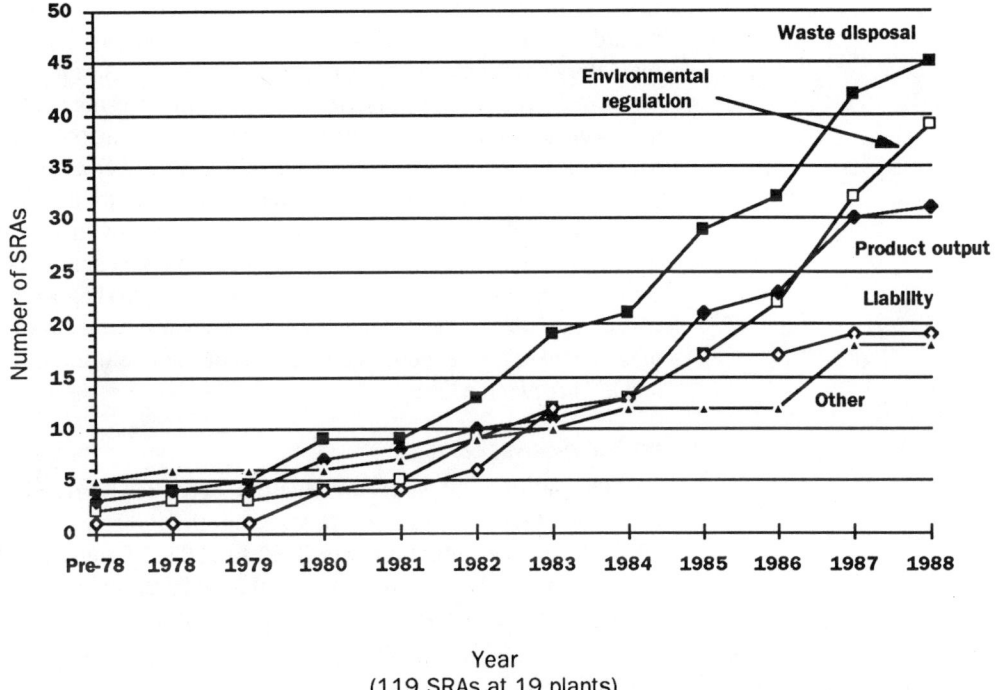

Year
(119 SRAs at 19 plants)

Environmental regulations

Environmental regulations have been the fastest growing incentive in recent years, cited as the reason for 40 source reduction activities implemented between 1985 and 1990, as opposed to 10 activities implemented before 1985 — a 300 percent increase. Figure 16 shows that environmental regulations, as a motivating factor for source reduction activities, began to rise dramatically in 1984. Environmental regulations at that time did not directly require source reduction action, but included requirements such as stricter limitations on transfers of hazardous or toxic waste to publicly owned treatment works (POTW).

It is interesting to note that, although the rise began in 1984, none of the INFORM study plants specifically mentioned the 1984 Hazardous and Solid Waste Amendments (HSWA) to the Resource Conservation and Recovery Act (RCRA) as a reason for implementing a source reduction activity. While these amendments contained the first provisions by the federal government calling for preventive measures, they did not place very stringent demands on industry. Rather, they simply required that generators of solid hazardous waste certify that a program was in place to minimize waste generation. At that time, the term "minimization" was a "catch-all" term including every means of reducing solid waste disposal, even waste treatment.

Only one plant, Fisher Scientific in Fair Lawn, New Jersey, specifically cited source reduction legislation as an incentive to seek source reduction opportunities. Fisher told INFORM that debate in New Jersey over the state's Pollution Prevention Act (subsequently passed in the summer of 1991) spurred the plant not only to determine how it could best use end-of-pipe controls to reduce chemical releases but also to take a closer look at the materials it uses and its reasons for using them. According to plant officials, the discussion in the state "started the challenge" and the company "got some interesting answers."

In some cases, existing environmental regulations discourage source reduction. Two examples of these, discussed in depth in *Cutting Chemical Wastes*,[1] were still in place at the time of INFORM's follow-up research for this report: regulations affecting burning of hazardous wastes as an alternate fuel and deep well injection of hazardous waste. Companies burn waste as fuel because they can save money on energy costs and on hazardous waste management: wastes burned as fuel are exempt from the stringent RCRA regulatory requirements that apply to these same wastes if they are burned in hazardous waste incinerators. Similarly, deep well injection has not been strictly regulated and thus has been a relatively inexpensive option for the disposal of hazardous wastes. The accessibility of inexpensive waste disposal options such as these creates little incentive for plant managers to search for source reduction opportunities.

Additionally, while in some cases the stringent laws in California and New Jersey may have caused plant managers to take source reduction actions, while no comparable laws existed in Ohio, overall the amount of source reduction happening at INFORM's study plants was independent of the differences in state environmental laws. There were an average of 5.3 reported source reduction activities per plant in California, 10.2 per plant in New Jersey, and 8.7 per plant in Ohio.

Waste disposal costs

The number of source reduction activities spurred by waste disposal costs increased steadily and steeply throughout the 1980s, with a 119 percent increase from 1985 (16 source reduction activities) to 1990 (35 source reduction activities). This pattern appears likely to continue into the 1990s as well, as waste disposal costs continue to rise and disposal options continue to decrease.

Product output

Source reduction activities spurred by product output issues began to rise sharply in 1984, but leveled off after 1987, for a 111 percent increase from 1985 (9 source reduction activities) to 1990 (19 source reduction activities). Product output issues include operating costs, product yield and product quality improvements, and customer specifications.

Liability

Issues of liability as a factor motivating source reduction activities increased by 50 percent between 1985 (12 source reduction activities) and 1990 (18 source reduction activities), but a sharp increase occurred in 1979 (when the Superfund law, with its joint and several liability provisions for waste generators, was being debated in Congress) and another sharp increase took place in 1982. Liability issues also include worker safety and community relations.

Other

Source reduction motivated by "other" factors has not exhibited any tendency to greatly spur the number of source reduction actions taken (only 7 source reduction activities in 1990) and increased only 17 percent from 1985 (6 source reduction activities) to 1990. Other motivations include such factors as "self-initiated review" (as used on the TRI form R), lowering energy costs, and employee suggestions. Employee suggestions were cited as the motivating factor for 9 source reduction activities. However, there is no way of knowing how many other source reduction activities (motivated by cost, regulations, etc.) also involved employee suggestions in identification of the particular opportunity.

---

1 INFORM, *Cutting Chemical Wastes*, pp. 121-125.

Motivation by type of waste reduced

Motivations varied depending on the type of waste reduced: disposal issues were the most frequently cited incentive for reducing solid wastes (55 percent), as they were overall, but regulations were the most frequently cited reason for activities reducing air emissions (47 percent) and wastewater discharges (39 percent). Table I-10 lists the reasons cited for implementing a source reduction activity for each type of waste, and Figure 17 illustrates these figures as percentages of the total number of source reduction activities (thus, the percentages for each type of waste may total more than 100 percent since more than one reason may be given for undertaking a given source reduction activity).

Table I-10: Number of Source Reduction Activities by Motivating Factor and Type of Waste

	Disposal Costs	Environmental Regulation	Product Output	Liability	Other
Air emissions	8	16	10	2	7
Wastewater	23	27	19	12	7
Solid wastes	35	18	18	15	8

Figure 17: Motivation for Implementing Source Reduction Activities (SRAs) by Type of Waste*

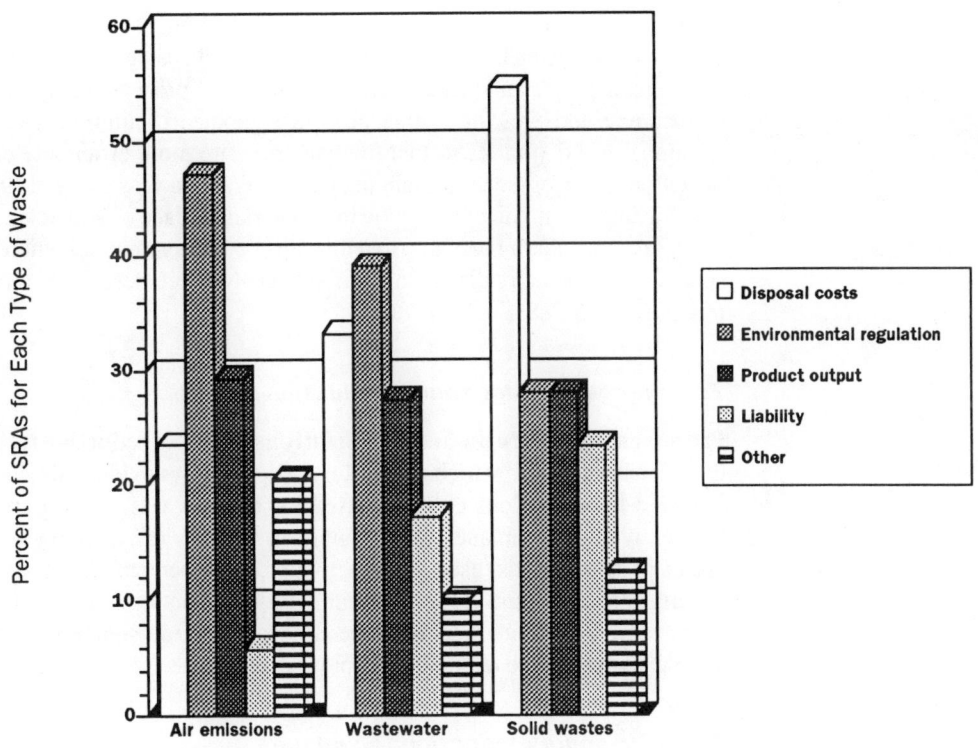

Type of Waste
(142 SRAs at 21 plants)

* *Within each category of wastes, the percentages total more than 100 percent since one source reduction activity can be motivated by more than one motivating factor.*

Air emissions

For air emissions, regulations were cited as a motivating factor for 47 percent of the source reduction activities, followed by product output (29 percent), waste disposal costs (24 percent), "other" factors (21 percent), and liability concerns (6 percent). Air emissions have generally been less regulated than wastewaters and solid wastes, with air permit requirements one of the few environmental regulations affecting air emissions.

Wastewaters

For wastewaters, regulations were cited as a motivating factor for 39 percent of the source reduction activities, followed by waste disposal costs (33 percent), product output (28 percent), liability (17 percent), and "other" reasons (10 percent). Strict limitations on quantities of wastes that can be transferred to publicly owned treatment works are one source of regulations affecting wastewater discharges.

Solid wastes

For solid wastes, waste disposal costs played the most significant role in motivating source reduction activities (55 percent), with regulations and product output (28 percent each), liability (23 percent), and "other" reasons (13 percent) playing lesser roles. It is worth noting that waste disposal costs are affected by hazardous waste regulations, through their focus on waste treatment facilities.

Source Reduction Techniques

Table I-13, starting on page 52 at the end of this section, lists the 137 source reduction activities INFORM documented in the second round of research, categorized under the five major types of source reduction techniques described in Chapter 1: process changes (such as better control of temperature and pressure within process equipment), equipment changes (such as adding additional reaction tanks so that raw materials are more efficiently converted into product), operations changes (such as changing the way reaction vessels are cleaned), chemical substitutions (using a nontoxic or less toxic raw material in place of a toxic one), and product changes (such as reformulating paints so that they do not require toxic solvent bases). The remaining 44 source reduction activities included in this analysis are described in depth in INFORM's 1985 report, *Cutting Chemical Wastes*.

Techniques used for source reduction activities

Process changes were the most frequently used source reduction technique (used for 78 out of 177 activities for which this information was provided, or 44 percent of the total), followed by operations changes (used for 60 activities, or 34 percent of the total). Less frequently cited techniques were equipment changes (used for 31 activities, or 18 percent), chemical substitutions (used for 18 activities, or 10 percent), and product changes (used for 8 activities, or 5 percent). (The total number of techniques adds up to more than 177, and the percentages total more than 100 percent, because some source reduction activities involved more than one source reduction technique.)

Trends in source reduction techniques

Both process changes and operations changes consistently increased during the 1980s, and are continuing to do so, while product reformulations have increased only minimally since 1978 and consistently are the least used source reduction technique. Equipment changes have increased more slowly, with fewer new ones introduced each year. Chemical substitutions increased dramatically between 1986 and 1987. In 1986, Congress passed the Emergency

Planning and Community Right-to-Know Act (also know as Title III of the Superfund Amendments and Reauthorization Act, or SARA) which mandated the Environmental Protection Agency to require industry to annually report releases and transfers of specific toxic chemicals used and/or manufactured, beginning with 1987 data; however, few new chemical substitutions were used in 1988.

Figure 18 shows the total number of source reduction activities in place each year (those implemented within the year as well as those implemented in previous years) for each source reduction technique used. By showing trends over time, this figure illustrates that plants are continuing to identify efficiency-oriented source reduction techniques (process, operations, and equipment changes), while the more innovative techniques (chemical substitutions and product changes) continue to lag behind. (The numbers of source reduction activities for each technique shown in this figure differ from the numbers discussed in the previous section because the figures are based on only those source reduction activities for which information on the year of implementation was also provided.)

Opportunities for chemical substitution in the chemical manufacturing industry exist in both the production process itself (9 of the 11 chemical substitutions for which this information was provided) and in nonproduction functions (2 of the 11).

Source reduction activities focused on solvents can be effective in reducing hazardous and toxic wastes both because toxic and hazardous chemicals can be used as solvents and because, since solvents are not incorporated into products, they can often end up in plant waste. At least four chemical substitutions (and possibly two more), out of 13 for which this information was available, involved solvents. Information on the remaining chemical substitutions was inadequate for this analysis.

Figure 18: Cumulative Number of Source Reduction Activities (SRAs) per Year by Source Reduction Technique Used

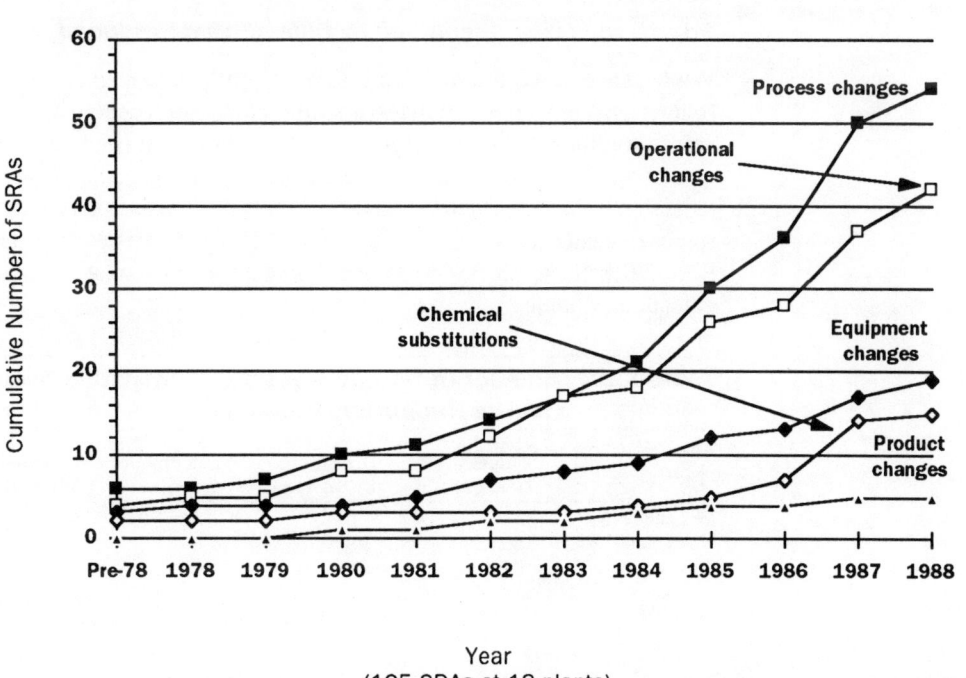

Year
(125 SRAs at 18 plants)

Amount of waste reduced, by source reduction technique

Source reduction techniques contributing to increased efficiency of existing processes (process, operations, and equipment changes) each reduced individual wastestreams, on average, by about 70 percent. Chemical substitutions reduced the target wastestream, on average, by about 50 percent. While there were only three uses of product changes for which data on the amount of waste reduced were provided, each completely eliminated the targeted wastestream. Table I-11 shows the average percent reduction achieved for each source reduction technique used.

Table I-11: Average Percent Wastestream Reduction by Source Reduction Technique Used*

Technique Used	Average Percent Wastestream Reduced per Source Reduction Activity
Process changes	72%
Operations changes	70%
Equipment changes	68%
Chemical substitutions	48%
Product changes	100%
Combinations	84%

* Based on 96 source reduction activities at 20 plants.

Chemical substitutions may have showed a low percent reduction because substitutions were not always made to nontoxic or nonhazardous materials. For example, while Chevron Chemicals in Richmond, California, implemented at least three chemical substitutions, the company reported that the substitutes as well as the original chemicals are considered hazardous. (The substitutions allowed Chevron to use less raw materials to produce the same amount of product.)

Source reduction technique by type of waste reduced

While process changes were the most frequently used source reduction technique overall, followed by operations changes, equipment changes, chemical substitutions, and product reformulations, the percentages varied depending on the type of waste reduced. For each type of waste, Table I-12 shows the number of source reduction activities using each source reduction technique. Figure 19 illustrates these figures as percentages of the total number of source reduction activities. (The percentages for each type of waste in Figure 19 may total more than 100 percent, since one source reduction activity may involve more than one source reduction technique.)

Table I-12: Number of Source Reduction Activities by Type of Waste and Source Reduction Technique

	Process Changes	Operations Changes	Equipment Changes	Chemical Substitutions	Product Changes
Air emissions	14	8	17	1	1
Wastewater	36	29	6	10	1
Solid waste	31	24	9	5	5

Air emissions

For air emissions, equipment changes (such as adding condensers to trap and return fugitive emissions) were used for 49 percent of the source reduction activities, followed by process changes (40 percent), operations changes (23 percent), and chemical substitutions and product reformulations (3 percent each).

Wastewaters

For wastewaters, process changes were used for 49 percent of the source reduction activities, followed by operations changes (40 percent), chemical substitutions (14 percent), equipment changes (8 percent), and product changes (1 percent).

Solid wastes

For solid wastes, process changes were used for 45 percent of the source reduction activities, followed by operations changes (35 percent), equipment changes (13 percent), and chemical substitutions and product reformulations (7 percent each).

Figure 19: Techniques Used for Source Reduction Activities (SRAs) by Type of Waste*

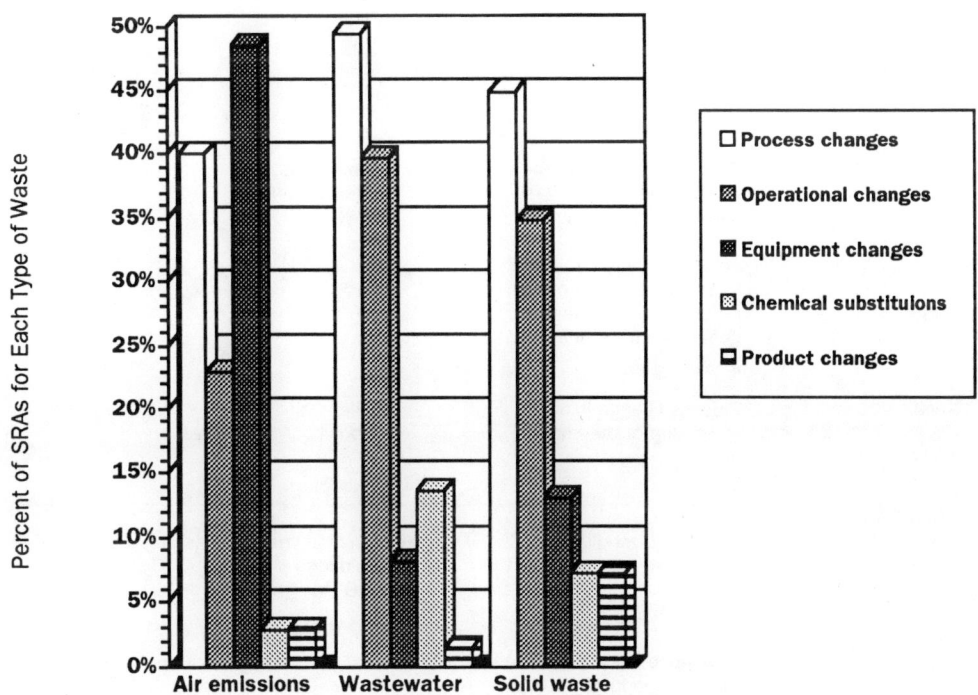

Type of Waste
(151 SRAs at 21 plants)

* *Within each category of wastes, the percentages may total more than 100 percent since one source reduction activity may involve more than one source reduction technique.*

Table I-13:	137 Source Reduction Activities Categorized by Technique Used			
Waste Medium (SR Type) Year	Source Reduction Activity	Specific Waste Reduced (Hazardous or Nonhazardous)	Percent Waste Reduced	Amount Waste Reduced

PROCESS CHANGES

American Cyanamid, OH				
Solid/water (PS,CH) 1987	Modified a yellow dye manufacturing process to substitute new solvent for nitrobenzene. New solvent recovered through in-process recycling.	Nitrobenzene waste (H)	100%	200,000 lb/yr
		Biological oxygen demand (BOD) (N)	90%	120,000 lb/yr
Air (PS) 1980+	Switched from producing dry products to producing mostly wet products.	Dust (N)	90%	2,000 lb/yr
Aristech, OH				
Solid (PS) 1984	Recycle bisphenol-A (BPA) mother liquor before last step in the process.	Heavy ends, organic waste (H, N)		
Solid (PS) 1985	Process changes resulted in a purer grade of phenol reducing organic waste generation from BPA production.	Heavy ends (H, N)		
Solid (PS) 1987	Process change in the wash step of the Phenol I unit.	Boiler slag deposits (H)	25%	
Atlantic, NJ				
Water (PS) 1984-1987	Changed crystallization time for some products.	Reaction vessel residues BOD (N)		
Water (PS) 1984-1987	Developed liquid processes that are conducted in-situ. Dyes are produced in a single kettle with no need to filter or rinse.			
Water (PS) 1984-1987	Searching for ways to increase concentration of final product so it can be spray dried instead of precipitated, isolated, and filtered.			
Water (PS) 1984-1987	Changed relative concentrations of reagents in order to drive reactions further to completion and increase yields.			

Key to source reduction types: CH, chemical substitution; EQ, equipment change; OP, operational change; PR, product change; PS, process change.
A blank indicates that the plant did not provide information.

Change in Yield	Dollars Saved	Dollars Spent	Motivation	Comments	Time Needed for Implementation
None	$200,000/yr	$100,000 for equipment	Waste cake banned from landfilling; costs and problems of incineration. New limitations of BOD at city sewage treatment plant.	Cost savings from reduced disposal costs and energy savings from not having to recover nitrobenzene.	Several months
Unknown	Unknown	$0	Customers preferred wet products.		2 yr for one major product line
	$50,000/yr		Result of employee suggestion.	Savings from increased yield.	
			Process changes identified through the Aristech Total Performance (ATP) program.		
			Process change identified through the ATP program.	Process change improves yield, reducing boiler slag deposits when heavy ends are burned.	
			Increase yields and improve productivity.	Fewer rinses reduces washings and lost product; 50% reduction in water use.	
	10% or more of manufacturing costs		Increase yields and improve productivity.	Overall impact small; just 1 or 2% of production processes have been affected.	
			Reduce energy costs.	Wastewater as well as energy costs would be reduced.	
			Increase yields and improve productivity.		

(continued)

Table I-13:	137 Source Reduction Activities Categorized by Technique Used (continued)			
Waste Medium (SR Type) Year	Source Reduction Activity	Specific Waste Reduced (Hazardous or Nonhazardous)	Percent Waste Reduced	Amount Waste Reduced

PROCESS CHANGES (cont'd.)

Waste Medium (SR Type) Year	Source Reduction Activity	Specific Waste Reduced	Percent Waste Reduced	Amount Waste Reduced
Chevron, CA				
Air (EQ, PS, OP) 1987-1990	Process modifications reduced fugitive emissions of dichloromethane. Also, double-barrier seal pumps were added to detect leaks and allow switch-over to other pumps while leaks are repaired.	Dichloromethane (H)	78%	140,000 lb
Water (CH,PS) 1989		(H)	50%	
Ciba-Geigy, NJ				
Solid/water (PS) 1985	Excess metals eliminated in dye manufacture by shifting purification to early part of process and improving process controls.	Heavy metals (H)	100% for six products	100 drums (45,000 lb/yr)
Solid/water (PS) 1986	Use purified intermediates so can reuse metallization filtrate when an excess of metal is needed to complete the reaction.	Heavy metals (H)	80%	40 drums (18,000 lb/yr)
Solid/water (PS,OH) 1986	Process change to eliminate nitro separation step and replace iron as raw material.	Iron sludge (H)	100% of iron and 80% of total organic carbon (TOC)	
Water (PS) 1986	Process change to reduce amount of m-phenylene diamine used and discharged in filtrate water.	Amino coupling component (H)	80% of TOC in filtrate	
Solid (PS) 1987	A new antioxidant process eliminates the need for excess hydrazine.	Hydrazine precipitate (H)	100%	90,000 lb/yr
Air (PS,EQ)	Use pumps to transfer liquids rather than blowing with nitrogen.			
Air (PS)	Installed computer control for maintenance of inert nitrogen atmosphere.			
Def-Tec, OH				
Water/solid (PS)	Def-Tec changed its product labeling operations from paint on paper to direct ink application (through silk-screening).	Paint waste	88%	253.5 gal/yr (2,535 lb/yr)

Key to source reduction types: CH, chemical substitution; EQ, equipment change; OP, operational change; PR, product change; PS, process change.
A blank indicates that the plant did not provide information.

Change in Yield	Dollars Saved	Dollars Spent	Motivation	Comments	Time Needed for Implementation
		$200,000	Employee suggestion. Dichloromethane has been reviewed as an air contaminant since the early 1980s.	Dollars spent is an estimate for new pumps only.	
			Employee suggestion.	The original two chemicals and their substitutes are both considered hazardous. This source reduction paid for itself in less than 6 months.	
+11%	$86,500	$0	New ocean discharge permit standards and high cost of disposal.		2 mo
+10%	$11,600	$0	Eliminate toxic components to meet new ocean discharge permit standards.		1 mo
+40%	$740,000	$0	Eliminate toxic components to meet new ocean discharge permit standards.	Reduce loss of product in wastewater and need for filtration.	9 mo
+12.5%	$250,000	$0	Eliminate toxic components to meet new ocean discharge permit standards.	Five percent reduction in use of m-phenylene diamine in the process.	3 mo
	$335,000	$200,000	Increased costs to recycle and recover raw material.	Recovery had been done by supplier that did not want to register as a treatment, storage, and disposal (TSD) facility.	2 yr
				Minimize nitrogen flow and, thus, entrainment of volatile materials.	
		$10,000 for silk-screening equipment	Saves operating costs and customers prefer the new label.	Savings from reduced cleaning operations, purchases of paint, and administrative time used for keeping track of labels.	

(continued)

Table I-13: **137 Source Reduction Activities Categorized by Technique Used** (continued)

Waste Medium (SR Type) Year	Source Reduction Activity	Specific Waste Reduced (Hazardous or Nonhazardous)	Percent Waste Reduced	Amount Waste Reduced
PROCESS CHANGES (cont'd.)				
Dow, CA Water/air (PS) 1987	A process change to an acid gas adsorption system where the wastestream is first scrubbed with water and then caustic (rather than just caustic) to avoid formation of compounds which would preclude recycling the spent caustic.	Spent caustic HCl (H)	100% of wastewater	500 tons/mo (1,000,000 lb/mo or 12,000,000 lb/yr) 160,000 lb/yr
Du Pont, NJ Water (PS) early 1960s	Butanol used as a solvent in oxyamines building is closed-loop recovered and reused.	1-Butanol (H)		10,000,000 lb/yr
Solid (PS) 1980	Still bottoms from oxyamines building are used as a raw material in another product.			75,000 lb/yr
Water (PS) pre-1970	Methanol, a solvent used in dimethylaniline building, is recovered and reused.	Methanol (H)		4,000,000 lb/yr
Water (PS) 1985	In the specialty intermediates building, cleanout solvent recovery is used to recover ortho-dichlorobenzene.	ortho-Dichlorobenzene (H)		750,000 lb/yr
Solid (PS) 1985	Still bottom purge streams are redistilled from the chloroamines process and recycled.			22,000,000 lb/yr
Solid (PS) 1986	Off-quality parachloroaniline flakes are distilled and reused in the same process.	Parachloroaniline flakes (H)		65,000 lb/yr
Water (PS) 1988	(1) Process improvements reduced use of methyl ethyl ketone (MEK); (2) waste MEK recycled back to process.	MEK (H)	50%	(1) 200,000 lb/yr reduced (2) 250,000 lb/yr recycled to process
Water (PS) 1979-1982	Process control improvements in Monastral process reduced use of nitrochlorobenzene.	Nitrochlorobenzene	65%	750,000 lb/yr
Exxon, NJ Solid (PS) 1984	Continuous process optimization of operating conditions has resulted in reducing generation of acid coke, a process residue.	Acid coke residue (H)	90%	157 tons/yr (314,000 lb/yr)
Solid (PS) 1983	Process optimization has reduced waste catalyst from a batch alkylation reactor.	Catalyst (H)	70%	11 tons/yr (22,000 lb/yr)

Key to source reduction types: CH, chemical substitution; EQ, equipment change; OP, operational change; PR, product change; PS, process change.
A blank indicates that the plant did not provide information.

Change in Yield	Dollars Saved	Dollars Spent	Motivation	Comments	Time Needed for Implementation
	$2,400,000/yr	$250,000	Closing of evaporation ponds.	Impurities are not produced in the acid neutralization step, so the brine can be reused to produce chlorine.	4 mo
	$2,750,000/yr		Cost reduction		
			Increased costs of landfilling and process yield improvement.		
	$350,000/yr		Operating costs reduction.		
	$260,000/yr		Operating costs reduction.		
			As part of program to reduce landfilled waste banned in New Jersey.		
			New Jersey banned landfilling of this material.	Minor capital expenses mainly for equipment to unload drums.	
None	$120,000/yr		Operating cost reduction (raw material costs plus reduced cost of incineration).		
			Operating cost reduction.		
	$340,000/yr		Undertaken to reduce potential long-term liability and disposal costs of waste previously landfilled. Remaining waste now incinerated.	Waste per unit of production dropped from 83 to 7. Savings are in incineration costs.	5 yr
	$14,000/yr	$0	Undertaken to reduce downtime for catalyst changes and to reduce disposal costs and long-term liability. Current disposal costs are $9,000 for 7 tons.	Waste per unit of production dropped from 19 in 1983 to 6 in 1986. Savings are in incineration costs.	

(continued)

Table I-13: 137 Source Reduction Activities Categorized by Technique Used (continued)

Waste Medium (SR Type) Year	Source Reduction Activity	Specific Waste Reduced (Hazardous or Nonhazardous)	Percent Waste Reduced	Amount Waste Reduced
PROCESS CHANGES (cont'd.)				
Exxon, NJ (cont'd.) Solid (PS,CH) 1984	A hydrocarbon raw material replaces oil as the phenol-absorbing medium in a manufacturing unit. The hydrocarbon and phenol mixture in the blowdown tank is recycled as feed to the unit.	Waste oil containing phenols (H)	100%	240 tons/yr (480,000 lb/yr)
Solid (PS) 1972	Replaced filters with high-speed centrifuges that remove the solids from lubricating oil additives without the addition of filter aid. Second-stage separation devices installed to recover the oil and active ingredients remaining in the centrifuge sludge.	Filter cake process solids (N)	68%	4,000 tons/yr (8,000,000 lb/yr)
Fisher, NJ Solid (PS)	Reintroduce impure forecut (first material distilled) and tailcuts (tail-end materials) into the process.			
Water/air (PS)	Two-step production of acetonitrile changed to a single-step distillation process, eliminating need to rinse a second reactor vessel and resulting air emissions.			
(PS) 1987	Improvements in quality control during operations led to decrease in waste generated from rejected product.			
(PS) 1990	Reducing amount of reactant raw materials in purifying tetrahydrofuran decreases off-specification solvent.	Solvent		30 gal/batch (2,200 lb/yr)
(PS) 1989-1990	Redesign of packaging line and filler hopper and use of lower-grade material has reduced need to flush solvent filling machine.	Solvent		90,000 lb/yr
Solid (PS) 1990	Improved in-process monitoring of Freon has increased yields and quality of product, reducing low-grade waste material.	Freon (H)		15,000 lb/yr
(PS) 1990	Improved in-process monitoring of quality of product (methanol) reduces number of times column material used to absorb impurities is changed.	Slurry and methanol (H)		6,000 lb/yr
(PS) 1985	Process changed to eliminate use of oleum (fuming sulfuric acid) as a solvent in processing acetonitrile.	Fuming sulfuric acid (H)	100%	

Key to source reduction types: CH, chemical substitution; EQ, equipment change; OP, operational change; PR, product change; PS, process change.
A blank indicates that the plant did not provide information.

Change in Yield	Dollars Saved	Dollars Spent	Motivation	Comments	Time Needed for Implementation
	$83,000/yr	$0	Undertaken to reduce waste disposed of off-site and to increase raw material savings.	Savings are in disposal costs.	
	$1,560,000/yr	$18,700,000	The large volumes of solid waste mixed with product made it obvious that there were opportunities to reduce product losses and disposal costs.	Waste per unit of product reduced from 105 in 1980 to 34 in 1986. Remaining waste landfilled.	
			New management investigation of ways to improve yields.	Forecuts and tailcuts had been sold as fuel or otherwise diposed of.	
			Improving yield of the process (one of highest volume products) and eliminating the need for three pieces of equipment.	Technical assistance was received from Du Pont.	18 mo
	$490,000/yr		Concern for worker safety and community relations; reduced disposal costs.		
			Concern for worker safety and community relations; reduced disposal costs.		
			Concern for worker safety and community relations; reduced disposal costs.		
			Concern for worker safety and community relations; reduced disposal costs.		
			Concern for worker safety and community relations; reduced disposal costs.		
			Concern for worker safety and community relations; reduced disposal costs.		

(continued)

Table I-13: 137 Source Reduction Activities Categorized by Technique Used (continued)

Waste Medium (SR Type) Year	Source Reduction Activity	Specific Waste Reduced (Hazardous or Nonhazardous)	Percent Waste Reduced	Amount Waste Reduced
PROCESS CHANGES (cont'd.)				
Fisher, NJ (cont'd.) (PS) 1987	Process changed to eliminate use of oleum (fuming sulfuric acid) as a solvent in processing hexane.	Fuming sulfuric acid (H)	100%	
(PS) 1989	Process changed to eliminate use of oleum (fuming sulfuric acid) as a solvent in processing iso-octane.	Fuming sulfuric acid (H)		
(PS) 1990	Process changed to reduce use of oleum (fuming sulfuric acid) as a solvent in processing cyclohexane.	Fuming sulfuric acid (H)	67%	
ICI Americas, CA Water (PS) 1987	Dewatering wastewater from chloro-propionamide production process yields trisodium phosphatedo-decahydrate, a commercially valuable material.	Trisodium phosphatedo-decahydrate (H)	81%	95,284 lb/yr
Water (PS) 1989	Raw materials in wastewaters reclaimed for reuse in production process of chloropropionamide.	(H)	19%	22,500 lb/yr
IFF, NJ (PS) 1987	Process change	Acetaldehyde (H)	34%	
Solid (PS) 1987	Process change	Chromium (H)	28%	
Air (PS) 1988	Changes in reaction conditions and reaction procedures allowed elimination of a solubilizer from one reaction step.	Isopropyl alcohol (H)	100%	
Air (PS) 1988	Use of different catalyst system and reaction conditions improved yields and eliminated organic chlorides as by-product.	Organic chlorides (H)	100%	
Monsanto, OH Solid (PS, PR) 1984-1985	The phenol-formaldehyde resins unit was replaced by a methylated melamine-formaldehyde resins process.	Phenol-formal-dehyde resin (H)	89%	16 drums/mo average (28,800 lb/yr)
Water (PS) 1987	The plant has upgraded the Scripset resin filter press to reduce leaks from the press.	Scripset press leakage (H)	96%	52 drums/mo (249,600 lb/yr)

Key to source reduction types: CH, chemical substitution; EQ, equipment change; OP, operational change; PR, product change; PS, process change.
A blank indicates that the plant did not provide information.

Change in Yield	Dollars Saved	Dollars Spent	Motivation	Comments	Time Needed for Implementation
			Concern for worker safety and community relations; reduced disposal costs.		
			Concern for worker safety and community relations; reduced disposal costs.		
			Concern for worker safety and community relations; reduced disposal costs.		
			Self-initiated review.	Overall waste generation increased by 26% but production increased by 60%.	
			Disposal costs.	Overall waste generation increased by 12% but production increased by 40%.	
			Air permit conditions.		4 mo
			Air permit conditions.		3 mo
	$4,800/mo		Market demand for new products. Older processes cannot compete on a production cost basis.	The new resins process is much cleaner, generating only two drums of hazardous waste per month.	
	$300,000/yr	$60,000	Process optimization.	Savings are from avoided incineration costs.	None

(continued)

Table I-13: 137 Source Reduction Activities Categorized by Technique Used (continued)

Waste Medium (SR Type) Year	Source Reduction Activity	Specific Waste Reduced (Hazardous or Nonhazardous)	Percent Waste Reduced	Amount Waste Reduced
PROCESS CHANGES (cont'd.)				
Monsanto, OH (cont'd.)				
Water (PS) 1986-1987	Statistical process control measures were applied to the ABS process to chart key indicators, including solids in wastewater.	Polymer solids		
Solid (PS) 1972-1987	Process changes to enable additional recovery and sale of spent monomers.	Monomers (H)	50%	1,400,000 lb/yr
Solid (PS) 1986	The ABS/SAN compounding process was modified to reduce off-grade product.	ABS plastics		
Perstorp, OH				
Air (PS) 1988	Processes changes in the sodium formate process reduced amount of fine particles generated.	Dust (N)		
PMC, OH				
Water (PS) 1987	TCB had been washed out of the product with naphtha. The process was reworked so that the individual solvents can be recovered.	Trichlorobenzene (TCB) (H)	50%	5,000 lb/yr
Water (PS) 1990	Changes in process chemistry eliminated the need for TCB.	TCB (H)	100%	
Water (PS, CH) 1990	The process chemistry has been changed for three products so that non-regulated chemicals can be substituted for regulated ones.	Varied (H)	100%	
Rhône-Poulenc, NJ				
Solid/air (PS) 1982 or 1983	Reduction in VOS emissions achieved by eliminating use of a VOS.	Volatile organic solvent (VOS) (N)	100%	
Water (PS) 1987	Optimization of the salicylaldehyde process has resulted in yield improvements.	Salicylaldehyde and by-products (N)		60,000 lb/yr

Key to source reduction types: CH, chemical substitution; EQ, equipment change; OP, operational change; PR, product change; PS, process change.
A blank indicates that the plant did not provide information.

Change in Yield	Dollars Saved	Dollars Spent	Motivation	Comments	Time Needed for Implementation
+11.5%	$500,000-$1,000,000/yr	Negligible	Improved yields to reduce costs.		Less than 1 yr
			Continuous effort to upgrade processes.		20 yr & ongoing
+5%	$1,000,000/yr		Continuous effort to improve yields and reduce costs.		
			Improve yields and generate less dust.		
		$25,000	Pressure from overseas competitors forces PMC to look very carefully at losses due to waste generation.	Modifications were needed to prevent the generation of mixed solvents as well as the recovery of individual solvents.	6 mo
			Stricter limits on discharge to municipal sewage treatment plant.	This change resulted in total elimination of TCB at this facility.	1.5 yr
			Federal pretreatment regulations for organic chemical plants.	One product may have to be dropped to meet new requirements.	
	$45,000/yr in combination with equipment change.	$10,000 in combination with equipment change.	"Waste minimization" and cost reduction.		1 mo
+2%	$250,000/yr	$200,000	"Waste minimization" and cost reduction.		1 yr

(continued)

Table I-13: 137 Source Reduction Activities Categorized by Technique Used (continued)

Waste Medium (SR Type) Year	Source Reduction Activity	Specific Waste Reduced (Hazardous or Nonhazardous)	Percent Waste Reduced	Amount Waste Reduced
OPERATIONAL CHANGES				
American Cyanamid, OH				
Water (OP) 1988	Compartmentalized tank farms onto pads to contain spills, which are recovered for use when possible.	Spilled organic and inorganic materials (H)	90%	900 lb/yr
Water (OP) 1985	Recycle quality control samples from lab back to process wherever possible.	Quality control samples (H)	90%	9 drums/yr (2,700 lb/yr)
Aristech, OH				
Water (OP) 1987	Reused hazardous wastewater for making raw material solutions at Phenol I unit.	Wastewater (H)		1,717,000 gal/yr (14,279,000 lb/yr)
Water (OP) 1984	Drain liquid from spent bisphenol-A (BPA) process filters back into process instead of disposing as wastewater.	BPA (N)		85,000 lb/yr
Atlantic, NJ				
Water (OP) 1984-1987	Eliminated storage of oleum in tanks due to risk of spills. Oleum is now only used in drums.			
Solid (OP) 1984-1987	Changed from fiber drums to plastic drums which can be reused, for internal use.	Trash (N)	33%	
Borden, CA				
Water (OP, EQ) 1988	Recirculate formaldehyde-contaminated vacuum pump seal water back to process. Carbon steel pumps replaced with stainless steel pumps. Similarly, contaminated reaction seal water is recirculated in phenolic resin process.	Formaldehyde (H)	98%	6,000 lb/yr
Water (OP) 1987	Reuse flushings from truck loading filters. Filtrate from the first two flushes are returned to the process. The excess water is distilled off and reused in process.	Formaldehyde (H)	95% (30% chemical oxygen demand reduction for total plant)	150,000 lb/yr
Water (OP) 1987	Isolated the formaldehyde manufacturing unit with trenches and dikes to collect wastewater from leaks and equipment rinsing. Collected water is reused in process.	Formaldehyde (H)	100% (10% COD reduction for total plant)	50,000 lb/yr

Key to source reduction types: CH, chemical substitution; EQ, equipment change; OP, operational change; PR, product change; PS, process change.
A blank indicates that the plant did not provide information.

Change in Yield	Dollars Saved	Dollars Spent	Motivation	Comments	Time Needed for Implementation
	$10,000/yr	$700,000	Project to avoid treatment of on-ground spills in new on-site wastewater treatment plant.		6 mo for design
	Unknown	$0	Reduce toxicity and volume of wastewater stream to meet new requirements of the city's sewage treatment plant.		None
		$12,500	Result of employee suggestion.		
+85,000 lb/yr	$32,047/yr	$3,697	Result of employee suggestion.	Practice instituted after ensuring that all contact with air could be avoided so that the fluid would not solidify.	
			Concern that Atlantic live responsibly and at peace with the plant's residential neighborhood.		
			Quadrupling of cost of nonhazardous trash removal.	Overall costs have gone up due to increased removal costs.	
+0.005%	$420/yr	$20,000	Local sewage-treatment plant imposed a formaldehyde limit of 50 ppm.	Stainless steel pumps cost about $6,000 more than the carbon steel ones and do not corrode as concentration of formaldehyde increases.	0
+0.2%	$14,250/yr	$2,000	Local sewage-treatment plant imposed a formaldehyde limit of 50 ppm.	The main filter used in truck loading is commonly back-flushed three times a day.	1 person-month
+0.05%	$3,500/yr	$10,000 for pipes and tanks	Local sewage treatment plant imposed a formaldehyde limit of 50 ppm.		0

(continued)

Table I-13: 137 Source Reduction Activities Categorized by Technique Used (continued)

Waste Medium (SR Type) Year	Source Reduction Activity	Specific Waste Reduced (Hazardous or Nonhazardous)	Percent Waste Reduced	Amount Waste Reduced
OPERATIONAL CHANGES (cont'd.)				
Borden, CA (cont'd.)				
Water (OP) 1987	Modified product filter cleaning operation in urea-formaldehyde resin manufacturing process. Water is collected in Tote bins and reused in process.	Formaldehyde (H)	95% (15% COD reduction for total plant)	75,000 lb/yr
Solid (OP, EQ)	Reduced wastewater treatment sludge generation through formaldehyde-related activities.	Sludge (H)	80% reduction for total plant	
Water (OP) 1988	Collect and isolate wastewater in the trench coming from the tank house and phenol railcar unloading area in a 30,000 gallon tank. The water is reused in the resin batches.	Phenol (H)	100%	70 lb/yr
Chevron, CA				
Air (EQ, PS, OP) 1987-1990	A variety of small operational changes and process modifications reduced fugitive emissions of dichloromethane. Also, double-barrier seal pumps were added to detect leaks and allow switch-over to other pumps while leaks are repaired.	Dichloromethane (H)	78%	140,000 lb/yr
Water (CH,OP) 1970s (over 15-yr period)	Two chemical substitutions allow less raw material to be used per pound of product produced.	(H)	40%	
(OP)	Some quality assurance samples and all product quality samples are returned to the process.	(H)		
Ciba-Geigy, NJ				
Solid (OP) 1985	Use minimum quality control sample sizes and return samples to process.	Intermediate (N)	50%	20,000 lb/yr
Air (OP)	Limit open manhole operations.			

Key to source reduction types: CH, chemical substitution; EQ, equipment change; OP, operational change; PR, product change; PS, process change.
A blank indicates that the plant did not provide information.

Change in Yield	Dollars Saved	Dollars Spent	Motivation	Comments	Time Needed for Implementation
+0.1%	$7,200/yr	$4,000	Decision to discontinue underground storage tanks.	Particles are collected in drums for disposal.	1 person-month
	Additional $17,750/yr	$250,000	Rising costs of hazardous waste disposal.	Savings are in sludge removal costs and reduced sewer charges.	6 mo
Negative	$1,500/yr in treatment costs	$3,000	Corporate goal of zero discharge of phenolic wastewater.	Before, the wastewater went to an ozone treatment unit. Now this unit is only needed during the rainy season.	
		$200,000	Employee suggestion. Dichloromethane has been reviewed as an air contaminant since the early 1980s.	Dollars spent is an estimate for new pumps only.	
			Employee suggestion.	The original two chemicals and their substitutes are all considered hazardous.	
	$100,000	$0	Disposal cost savings.	Expense of handling of samples to reduce risk of cross-contamination is justified by rising disposal costs.	

(continued)

Table I-13: 137 Source Reduction Activities Categorized by Technique Used (continued)

Waste Medium (SR Type) Year	Source Reduction Activity	Specific Waste Reduced (Hazardous or Nonhazardous)	Percent Waste Reduced	Amount Waste Reduced
OPERATIONAL CHANGES (cont'd.)				
Colloids, CA				
Water/solid (OP)	Return samples of each product batch, as well as quality control samples, to the product batch.			
Solid (OP)	Use reusable Tote bins to supply product for one customer.			
Dow, CA				
Water (OP) 1987	Installed dikes to hold floods expected once in 100 years and material from the rupture of any one storage tank.			
Solid (OP) 1987	Survey of largest contributors to flow of whitewater eliminated many. A recycling system was installed to collect the remaining whitewater in a tank and return it to latex production process.	Latex solids	95%	
Du Pont, NJ				
Various (OP) 1983	Du Pont insists that suppliers supply high-quality raw materials.	Various (H, N)	Not quantified	
Exxon, NJ				
Solid (OP) 1980	Reduction in waste lubricating oil additives accomplished through better housekeeping practices and the scheduling of longer campaigns (larger volumes of an individual product are produced at one time, reducing the number of equipment washings needed).	Waste lubricating oil additives (H, N)	50%	3,796 tons/yr (7,592,000 lb/yr)
Fisher, NJ				
(OP)	Starting with higher-grade raw materials provides electronics industry customers with higher-grade products.			

Key to source reduction types: CH, chemical substitution; EQ, equipment change; OP, operational change; PR, product change; PS, process change.
A blank indicates that the plant did not provide information.

Change in Yield	Dollars Saved	Dollars Spent	Motivation	Comments	Time Needed for Implementation
			Customer requires that product be supplied in reusable Tote bins. Other customers do not buy enough of any one product to take advantage of the Tote bins.		
			Closing of evaporation ponds.	Total containment structures capture any spills/run-off from the process area for reuse as process water in the brine plant and chlorinolysis unit.	
	$26,000/yr plus $300,000 avoided capital investment in coagulation system		Avoid capital cost of new system and landfill disposal.	This avoided the need to coagulate the suspended latex particles and landfill them.	6 mo
			Higher product quality and lower waste generation.		
	$1,210,000/yr		Undertaken to reduce potential long-term liability when disposing of waste through a third party.	Waste lubricating oil additives are generated when lines and tanks are flushed. Savings are from reduced disposal costs and raw material purchases.	
			Electronics industry requires high-grade crystals and therefore, high-grade Fisher products.	With higher-grade raw materials, any rejected chemicals are of high enough grade to be sold to other customers.	

(continued)

Table I-13: 137 Source Reduction Activities Categorized by Technique Used (continued)

Waste Medium (SR Type) Year	Source Reduction Activity	Specific Waste Reduced (Hazardous or Nonhazardous)	Percent Waste Reduced	Amount Waste Reduced
OPERATIONAL CHANGES (cont'd.)				
Fisher, NJ (cont'd.)				
(OP)	Solvents are recovered from various phases of distillation process and reused in process or sold as product.	Solvents (H)	5%	1,500 gal/yr (9,350 lb/yr)
Water (OP)	Implementation of operator training to ensure minimal solvent losses and decrease cross-contamination of segregated solvents.			
Water (OP)	Solvents collected from routine line flushings are segregated for resale or reuse.		5%	3,500 gal/yr (21,830 lb/yr)
Solid (OP) 1987	Expired product and quality control samples are segregated, consolidated, and sold as product.		5%	1,000 gal/yr (6,250 lb/yr)
Solid (OP) 1989-1990	Flammable material from flushes and labs kept separate for sale as lower-grade product.	(H)		19,200 gal/yr (153,600 lb/yr or 13 drums)
Solid (OP) 1989-1990	Forecuts, tailcuts, and boilouts are analyzed for purity and potential resale as product.	(H)	60%	9,900 gal/yr (79,200 lb/yr)
(OP) 1990	Methanol contaminated with dolomite particles, left in still at end of batch when possible.	Methanol (H)		4,000 lb/yr
(OP) 1990	Before, for acids and ethers, a single bottle from a vendor's case was retained and the rest was sent for disposal. Now, either a single bottle is obtained or the partial cases are collected and sent to another plant for their inventory.	Acids/ethers		800 gal/yr acids and 400 gal/yr ethers
ICI Resins, CA				
Solid (OP)	The plant has instituted "ABC" inventory control so that stored materials do not go bad before they are used.			

Key to source reduction types: CH, chemical substitution; EQ, equipment change; OP, operational change; PR, product change; PS, process change.
A blank indicates that the plant did not provide information.

Change in Yield	Dollars Saved	Dollars Spent	Motivation	Comments	Time Needed for Implementation
	$2,250 in 1987	$0			
		$3,000		Dollar savings realized, but difficult to quantify.	
	$5,250 in 1987	$500			
	$1,500/yr	$500			
			Concern for worker safety and community relations; reduced disposal costs.		
			Concern for worker safety and community relations; reduced disposal costs.		
			Concern for worker safety and community relations; reduced disposal costs.		
			Concern for worker safety and community relations; reduced disposal costs.		
			System started to control costs of raw materials.	Inventory control leads to less spoilage of materials which then have to be disposed of.	

(continued)

Table I-13: 137 Source Reduction Activities Categorized by Technique Used (continued)

Waste Medium (SR Type) Year	Source Reduction Activity	Specific Waste Reduced (Hazardous or Nonhazardous)	Percent Waste Reduced	Amount Waste Reduced
OPERATIONAL CHANGES (cont'd.)				
ICI Resins, CA (cont'd.)				
(OP) 1985	Materials from leaks that cannot be eliminated are recycled. Catch pans are used underneath trucks during loading. Spills are put into drums along with lab samples and reused in the process.			
IFF, NJ				
Air/water (OP) 1988	Fine tuning of reaction temperature control improves yield. Fewer batches means decrease in wastewater from reaction cleaning and product purification.		7%	
Air (OP) 1988	Use of better heat-up control system on a special-purpose still results in fewer losses to still vacuum system.			
Monsanto, NJ				
Solid (OP) 1985	Improved operations prevents build-up of paraform wastes in tanks.	Paraform (H)	98%	103,000 lb over 2 yr
Solid (OP) 1986-1988	The plant has paved and isolated the area under each storage tank in order to keep spills from contaminating the ground.	Spills (H, N)		
Perstorp, OH				
Water (OP) 1987	Wastewater streams are automatically sampled for chemical oxygen demand (COD) every 3 minutes, 24 hours a day, to identify and correct problems as soon as they occur.	COD (N)	30%/unit of product	
Air (OP) 1986	Volatile organic compounds (VOCs) are monitored at 106 points throughout the plant.	VOCs (H)		
Water (OP) 1985	Improved housekeeping: spills are swept up, instead of rinsed away, and product is reused when possible.	Product (N)		

Key to source reduction types: CH, chemical substitution; EQ, equipment change; OP, operational change; PR, product change; PS, process change.
A blank indicates that the plant did not provide information.

Change in Yield	Dollars Saved	Dollars Spent	Motivation	Comments	Time Needed for Implementation
				All leak control measures combined account for 10 to 15% of reduced waste.	
+7%			Air permit conditions.		
			Air permit conditions.		
	$96,500 over 2 yr		Reduce waste generation and disposal costs.		
		$500,000	Environmental protection and reduction in clean-up costs.	If a spill does occur, the volume of contaminated soil to be cleaned up is not large and some spill material may be salvageable.	
	$40,000/yr		To improve yields.	Total COD in the wastewater has decreased even with an increase in production.	
			Regulations require monitoring.	Monitoring has made operators more aware of importance of maintenance to avoid leaks.	

(continued)

Table I-13: 137 Source Reduction Activities Categorized by Technique Used (continued)

Waste Medium (SR Type) Year	Source Reduction Activity	Specific Waste Reduced (Hazardous or Nonhazardous)	Percent Waste Reduced	Amount Waste Reduced
OPERATIONAL CHANGES (cont'd.)				
PMC, OH				
Water (OP) 1985	A discharge of concentrated bleach into a wastewater stream was eliminated by installing a continuous flow, closed-loop tank system. Bleach continuously flows through the loop and is sent to the reaction vessel (where it is completely reacted) only when needed.	Bleach (N)	100%	10,000 gal/yr (83,160 lb/yr)
Solid (OP) 1983	By replacing pump seals, the amount of soil contaminated with toluenediamine (TDA) due to leaks has been reduced.	TDA-contaminated soil (H)	59% from 1981 to 1987	2,350 lb/yr
Solid/Water (OP) 1982	By working with suppliers to provide higher-grade raw materials, significant yield increases and cost reductions have been realized.	Varied (N)		
Rhône-Poulenc, NJ				
Water (OP) 1987	All wastewater streams containing toluene are collected into one settling tank. The recovered toluene is put back into the processes through a closed-loop pipe system.	Toluene (H)	40%	100,000 lb/yr
Water (OP,EQ) 1990	Toluene will be collected in tank adjacent to operations unit, recovered through distillation, and reused in salicylaldehyde process.	Toluene (H)	20%	30,000 to 40,000 lb/yr
Solid (OP) 1970s	As standard practice, all quality control and raw material samples are sent back to be reused in the production processes.	Various (H, N)		3,000 lb/yr

Key to source reduction types: CH, chemical substitution; EQ, equipment change; OP, operational change; PR, product change; PS, process change.
A blank indicates that the plant did not provide information.

Change in Yield	Dollars Saved	Dollars Spent	Motivation	Comments	Time Needed for Implementation
			The change was made because of concern for the safety of workers who might come into contact with the concentrated bleach in the wastestream.	Bleach is cheap so costs savings are insignificant.	
	$10,000/yr		Safety concerns; and need to eliminate hazardous waste.	No cost savings because staff time had to be spent to find non-leaking pump seals for hot liquid.	
+10%	$250,000/yr		In the early 1980s, some yields had declined as much as 5% due both to process problems and raw material impurities.	Raw material purities have increased from 98 to 99.5%.	1 yr
	$16,000/yr	$40,000	Periodic process reviews, which identify potential for materials loss, noted that a lot of toluene was being discharged when all wastestreams were looked at.	Toluene is used in large quantities (about 3,500 gallons per day) as a solvent at the plant and losses should be small since it is not a reactant.	
	$4,500/yr	$4,000,000	Federal pretreatment regulations for organic chemical plants.		1 yr
	$20,000/yr		"Waste minimization" and cost reduction.		

(continued)

Table I-13: 137 Source Reduction Activities Categorized by Technique Used (continued)

Waste Medium (SR Type) Year	Source Reduction Activity	Specific Waste Reduced (Hazardous or Nonhazardous)	Percent Waste Reduced	Amount Waste Reduced
EQUIPMENT CHANGES				
Aristech, OH				
Solid (EQ) 1985 and 1987	Added additional reactor volume to the phenol and bisphenol-A (BPA) production units.	Heavy ends (H,N)		
Solid (EQ)	Installed new heavy ends tower with distillation column at Phenol I unit to increase product yield.	Heavy ends (H)	17%	9,450,000 lb/yr
Air/solid (EQ) 1988	Installed sample loops on product sampling purge lines to return the liquid to the BPA and phenol processes. A valve can be opened to obtain a sample when needed without exposing streams to the air.	BPA, phenol (N, H)		
Borden, CA				
Water (OP, EQ) 1988	Recirculate formaldehyde-contaminated vacuum pump seal water back to process. Carbon steel pumps replaced with stainless steel pumps. Similarly, contaminated reaction seal water is recirculated in phenolic resin process.	Formaldehyde (H)	98%	6,000 lb/yr
Solid (OP, EQ)	Reduced wastewater treatment sludge generation through formaldehyde-related activities.	Sludge (H)	80% reduction for total plant	
Solid (EQ) 1987	Package products in reusable Tote bins rather than nonreusable drums. Borden requires its suppliers to do the same.	Empty drums (H)	90%	300 drums/yr (12,000 lb/yr)
Chevron, CA				
Air (EQ, PS, OP) 1987-1990	A variety of small operational changes and process modifications reduced fugitive emissions of dichloromethane. Also, double-barrier seal pumps were added to detect leaks and allow switch-over to other pumps while leaks are repaired.	Dichloromethane (H)	78%	140,000 lb

Key to source reduction types: CH, chemical substitution; EQ, equipment change; OP, operational change; PR, product change; PS, process change.
A blank indicates that the plant did not provide information.

Change in Yield	Dollars Saved	Dollars Spent	Motivation	Comments	Time Needed for Implementation
	$2,400,000/yr	$7,071,000; expected payoff is 3 yr	From a program to increase yields through increased reactor capacity.	Increased residence time in reactor increases purity of product, reducing wastes.	
	$1,037,000/yr	$2,250,000		Engineering study showed that the heavy ends could be heated and cracked to produce more phenol.	
		$120,000	To comply with new volatile organic compound regulations of the Ohio EPA.	Purge lines used to be opened to the air and the samples disposed of off-site.	
+0.005%	$420/yr	$20,000	Local sewage treatment plant imposed a formaldehyde limit of 50 ppm.	Stainless steel pumps cost about $6,000 more than the carbon steel ones and do not corrode as concentration of formaldehyde increases. Three new pumps needed.	0
	Additional $17,750/yr	$250,000	Rising costs of hazardous waste disposal.	Savings are in sludge removal costs and reduced sewer charges.	6 mo
	$2,000/yr		Problems with drum disposal.	It is difficult to control use of the Tote bins and thus to clean them without generating wastewater, but they are popular because of the problem of disposing of drums.	
		$200,000	Employee suggestion. Dichloromethane has been reviewed as an air contaminant since the early 1980s.	Dollars spent is an estimate for new pumps only.	

(continued)

Table I-13: 137 Source Reduction Activities Categorized by Technique Used (continued)

Waste Medium (SR Type) Year	Source Reduction Activity	Specific Waste Reduced (Hazardous or Nonhazardous)	Percent Waste Reduced	Amount Waste Reduced
EQUIPMENT CHANGES (cont'd.)				
Chevron, CA (cont'd.) (EQ)	Automated quality assurance sampling equipment ensures that only a single vial of sample is delivered.	(H)		
Ciba-Geigy, NJ Solid (EQ) 1987	Installed separate dust collectors in new powder coatings process. Separates dust emissions so can be reused or sold as product.	Dust (N)		50 lb/batch (20,000 lb/yr)
Air (EQ) 1985	Designed sampling and charging devices for kettles in the resin solution production process to reduce solvent emissions.	Solvents (H)	90%	50 tons/yr (100,000 lb/yr)
Air (PS,EQ)	Use pumps to transfer liquids rather than blowing with nitrogen.			
Colloids, NJ Solid (EQ)	Installed a dust collector on a new dry blending tank. At the end of the run, collected material is emptied and put into a drum and sold.			
Def-Tec, OH Solid (EQ)	A consolidated press was replaced by a pellet-producing machine.			
Du Pont, NJ Water (EQ) 1986	An iodine recovery unit installed in the process. Recovered iodine is sold as product.	Iodine (N)	80 to 85%	200,000 lb/yr
Air (EQ) 1985	Compressors installed in tank trucks reduce vapor losses.	Various (H, N)	Not quantified	
Solid/water/air (EQ) 1982-1991	Wastewater collection system (closed, above-ground pipe) being built to replace open ditch system.			

Key to source reduction types: CH, chemical substitution; EQ, equipment change; OP, operational change; PR, product change; PS, process change.
A blank indicates that the plant did not provide information.

Change in Yield	Dollars Saved	Dollars Spent	Motivation	Comments	Time Needed for Implementation
+0.2%	$20,000/yr	$80,000	Cost of product and cost of disposal.	This process underwent the source reduction analysis required for all new products at Ciba-Geigy.	
+1%	$50,000/yr	$10,000	Regulatory compliance.		1 mo
			Best control equipment when new process equipment was purchased.		
		$17,000	Product yield increases.	The pellet producer generates less scrap and dust.	
	$275,000/yr	$1,000,000	Operating cost reduction.	Expect to pay off capital costs within 3 or 4 years.	
			Reduced air emissions and operating cost reduction.	Done as part of Du Pont's vent abatement program, result of mass balance calculations.	
		More than $10,000,000 (projected)	Environmental regulations.	Will eliminate evaporative losses and possible soil or groundwater contamination.	

(continued)

Table I-13: 137 Source Reduction Activities Categorized by Technique Used (continued)

Waste Medium (SR Type) Year	Source Reduction Activity	Specific Waste Reduced (Hazardous or Nonhazardous)	Percent Waste Reduced	Amount Waste Reduced
EQUIPMENT CHANGES (cont'd.)				
Fisher, NJ (EQ)		Solvents (H)	5%	20,000 gal/yr (124,740 lb/yr)
ICI Resins, CA (EQ)	Leaks have been reduced through the use of non-leaking pumps and replaced pipes and hoses.			
Merck, NJ Water/air (EQ) 1989	Installed process equipment to improve purification efficiency of Primaxin antibiotic production.	Methylene chloride (H)	50%	About 1,000,000 gal/yr (9,700,000 lb/yr)
Monsanto, OH Water (EQ) 1985	Replacement of phenol-formaldehyde resins unit by methylated melamine-formaldehyde resins process also has a methanol recovery distillation column that enables methanol in the wastestreams to be closed-loop recovered.	Methanol (H)	100%	15,600,000 lb/yr
Rhône-Poulenc, NJ Water (OP, EQ) 1990	Toluene will be collected in tank adjacent to operations unit, recovered through distillation, and reused in salicylaldehyde process.	Toluene (H)	20%	30,000 to 40,000 lb/yr
Air (EQ) 1987	In-line condensers, installed on the salicylaldehyde process, cool air lost during drying and recover the product from the emissions.	Salicylaldehyde (N)		10 lb/batch (average)
Solid/air (EQ) 1982 or 1983	Residues from the ethyl vanillin process were reduced by removing a piece of equipment that was degrading the product.	Ethyl vanillin by-products (N)		50,000 lb/yr

Key to source reduction types: CH, chemical substitution; EQ, equipment change; OP, operational change; PR, product change; PS, process change.
A blank indicates that the plant did not provide information.

Change in Yield	Dollars Saved	Dollars Spent	Motivation	Comments	Time Needed for Implementation
+28%	$30,000/yr	$75,000			
	About $1,000,000/yr	About $1,000,000	Plans to increase Primaxin production. Costs and risks of off-site hazardous waste treatment and transport.	Primaxin production doubled while solvent waste remains about the same.	
	$1,560,000/yr		Market demand.	Savings are for replacement material costs Waste is not generated due to process design.	
	$4,500/yr	$4,000,000	Federal pretreatment regulations for organic chemical plants.		1 yr
+0.5%	$30,000/yr	$10,000	To improve product yields. Product had been lost during the drying stage.	Odor reduction also achieved.	
	$45,000/yr in combination with process change.	$10,000 in combination with process change.	"Waste minimization" and cost reduction.		1 mo

(continued)

Table I-13: 137 Source Reduction Activities Categorized by Technique Used (continued)

Waste Medium (SR Type) Year	Source Reduction Activity	Specific Waste Reduced (Hazardous or Nonhazardous)	Percent Waste Reduced	Amount Waste Reduced
CHEMICAL SUBSTITUTION				
American Cyanamid, OH Solid/water (PS,CH) 1987	Modified a yellow dye manufacturing process to substitute new solvent for nitrobenzene. New solvent recovered through in-process recycling.	Nitrobenzene waste (H) Biological oxygen demand (BOD) (N)	100% 90%	200,000 lb/yr 120,000 lb/yr
Water (CH) 1985	Replaced a hazardous solvent, cellosolve acetate, with a nonhazardous solvent, ethylene glycoldiacetate.	Cellosolve acetate (H)	Unknown	
Aristech, OH Water (CH) 1989	Substituted a nonmetallic material for chromium used for corrosion resistance in cooling water.	Chromium (H)	100%	
Water (CH) 1988	Substituted Safety-kleen solvent for Dowclene.	Dowclene (H)	100%	
Chevron, CA Water (CH,OP) 1970s (over 15-yr period)	Two chemical substitutions allow less raw material to be used per pound of product produced.	(H)	40%	
Water (CH,PS) 1989		(H)	50%	
Ciba-Geigy, NJ Solid/water (PS,CH) 1986	Process change to eliminate nitro separation step and replace iron as raw material.	Iron sludge (H)	100% of iron and 80% of total organic carbon (TOC)	
Du Pont, NJ Solid (CH) 1986	Perlite replaced another filter aid, improving filter performance of wastewater treatment plant.	Filter aid (N)	50%	1,000,000 lb/yr
Exxon, NJ Solid (PS,CH) 1984	A hydrocarbon raw material replaces oil as the phenol-absorbing medium in a manufacturing unit. The hydrocarbon and phenol mixture in the blowdown tank is recycled as feed to the unit.	Waste oil containing phenols (H)	100%	240 tons/yr (480,000 lb/yr)
ICI Americas, CA Solid (CH) 1987	Switching to aqueous-based paints eliminated solvent wastes.	Solvent wastes (H)	100%	2,460 lb/yr

Key to source reduction types: CH, chemical substitution; EQ, equipment change; OP, operational change; PR, product change; PS, process change.
A blank indicates that the plant did not provide information.

Change in Yield	Dollars Saved	Dollars Spent	Motivation	Comments	Time Needed for Implementation
None	$200,000/yr	$100,000 for equipment	Waste cake banned from costs and problems of New limitations of BOD at sewage treatment plant.	Cost savings from reduced costs and energy savings from having to recover nitrobenzene.	Several months
	Unknown for customers, $10,000 at American Cyanamid	$0	Increases marketability of by eliminating hazardous label.	Change eliminates need for customer to dispose of any product as hazardous waste.	1 yr
			NPDES wastewater permit for chromium were tightened.	No cost savings. Increased water treatment costs by 40-50%.	
		$30,000 to ventilate work areas	Dowclene was classified as hazardous by the Ohio EPA to 1,1,1-trichloroethane.	Safety-kleen is flammable so areas had to be modified. It is recycled by vendor.	
			Employee suggestion.	The original two chemicals and substitutes are all considered hazardous.	
			Employee suggestion.	The original two chemicals and substitutes are all considered hazardous. This source reduction for itself in less than 6 months.	
+40%	$740,000	$0	Eliminate toxic components to meet new ocean discharge permit standards.	Reduced loss of product in water and need for filtration.	9 mo
			Operating cost savings plus improved filter performance.	Cost per pound of perlite is than former filter aid.	3 mo
	$83,000/yr	$0	Undertaken to reduce waste disposed of off-site and for raw material savings.	Savings are in disposal costs.	

(continued)

Table I-13: 137 Source Reduction Activities Categorized by Technique Used (continued)

Waste Medium (SR Type) Year	Source Reduction Activity	Specific Waste Reduced (Hazardous or Nonhazardous)	Percent Waste Reduced	Amount Waste Reduced
CHEMICAL SUBSTITUTION (cont'd.)				
IFF, NJ (CH) 1987	Substitution of chemical in process.	Ethylbenzene (H)	3%	
Air (CH) 1987	Substitution of chemical in process.	Hydrochloric acid (H)	4%	
(CH) 1987	Substitution of chemical in process.	Formaldehyde (H)	13%	
(CH) 1987	Substitution of chemical in process.	Toluene (H)	10%	
(CH) 1987	Substitution of chemical in process.	Xylene (H)	4%	
PMC, OH Water (PS, CH) 1990	The process chemistry has been changed for three products so that nonregulated chemicals can be substituted for regulated ones.	Varied (H)	100%	

Key to source reduction types: CH, chemical substitution; EQ, equipment change; OP, operational change; PR, product change; PS, process change.
A blank indicates that the plant did not provide information.

Change in Yield	Dollars Saved	Dollars Spent	Motivation	Comments	Time Needed for Implementation
			Self-initiated review.	Overall waste generation increased by 17% but production increased by 20%.	
			Self-initiated review.	Overall waste generation increased by 16% but production increased by 20%.	
			Self-initiated review.	Overall waste generation increased by 17% but production increased by 30%.	
			Disposal costs.	Overall waste generation increased by 10% but production increased by 20%.	
			Disposal costs.	Overall waste generation increased by 16% but production increased by 20%.	
			Federal pretreatment regulations for organic chemical plants.	One product may have to be dropped to meet new requirements.	

(continued)

Table I-13:	137 Source Reduction Activities Categorized by Technique Used (continued)			
Waste Medium (SR Type) Year	Source Reduction Activity	Specific Waste Reduced (Hazardous or Nonhazardous)	Percent Waste Reduced	Amount Waste Reduced

PRODUCT CHANGES

Waste Medium (SR Type) Year	Source Reduction Activity	Specific Waste Reduced (Hazardous or Nonhazardous)	Percent Waste Reduced	Amount Waste Reduced
American Cyanamid, NJ Solid (PR) 1987	Replaced 2 yellow dyes with another yellow dye using the same equipment.	6 waste cakes (N) Ammonia (H)	100%	500,000 lb/yr 100,000 lb/yr
Air (PS,PR) 1980+	Switched from producing dry products to producing mostly wet products.	Dust (N)	90%	2,000 lb/yr
Atlantic, NJ Water (PR) 1984-1987	Search for replacement products. Expect 2 or 3 approved under the pre-manufacture notification process of TSCA.			
Monsanto, OH Solid (PS,PR) 1984-1985	The phenol-formaldehyde resins unit was replaced by a methylated melamine-formaldehyde resins process.	Phenol-formaldehyde resin (H)	89%	16 drums/mo average (28,800 lb/yr)
Scher, NJ (PR) 1989	Eliminated two or three products that used hazardous chemicals as intermediates.	(H)	100%	

Key to source reduction types: CH, chemical substitution; EQ, equipment change; OP, operational change; PR, product change; PS, process change.
A blank indicates that the plant did not provide information.

Change in Yield	Dollars Saved	Dollars Spent	Motivation	Comments	Time Needed for Implementation
			Can no longer send certain wastes to the public sewage treatment plant.	Treatment and disposal not cost-effective.	6 mo
Unknown	Unknown	$0	Customers preferred wet products.		2 yr for one major product line
			Program to eliminate use of chemicals that could cause health or environmental problems.	Change likely to affect less than 1% of production. Costs could increase.	
	$4,800/mo		Market demand for new products. Older processes cannot compete on a production cost basis.	The new resins process is much cleaner, generating only two drums of hazardous waste per month.	
			Use of hazardous chemicals in their residential neighborhood not worth the risks.	Products were non-hazardous but chemicals used to make them were hazardous.	

Effect of Source Reduction on TRI Releases and Transfers

In the first round of case study research on these 29 chemical manufacturing plants, INFORM found that obtaining comprehensive and comparable waste generation data was nearly impossible. A major reason was that waste generation data collected by federal and state agencies was as varied as the number of statutes and agencies involved. Not only were different plants required to report on different types of chemical wastes, but information on the same chemical waste might be collected as pounds per hour for air emissions, as a concentration in wastewater, or as tons for solid waste. Further, where quantities were available for wastewater discharges and solid wastes, the amounts of inert materials such as water or soil were often included with, but not distinguished from, the amounts of toxic and hazardous constituents present.

The Toxics Release Inventory (TRI; see Chapter 1) has eliminated many of these differences and inconsistencies by requiring waste generators to report in a multimedia perspective on specific chemicals and chemical categories, to report on the constituent chemicals only (not on water or other inert materials), and to report the information in pounds. Such data begin to allow for a more accurate picture of a plant's overall waste generation involving these specific chemicals, as well as for a more meaningful comparison among plants.

The following analysis uses TRI data from the INFORM study plants to estimate the impact source reduction has had on the total amount of waste generated by these plants.

Total Toxics Release Inventory (TRI) releases and transfers from the chemical facilities INFORM studied would probably have been more than a third greater had it not been for achievements made in reducing wastes at the source.

The INFORM study plants reported that, as a result of 80 source reduction activities at 16 plants, 128.7 million pounds of waste per year are not now being generated. About 99 million pounds are wastes that were once generated but have been eliminated through the various types of source reduction studied in this report. The other 30 million pounds of this amount represent waste that was never generated. It was avoided by new processes designed to be less waste-producing than older ones in use at the time. For more than half of the source reduction activities (101 out of 181), no estimate was given for the amount of waste not generated (reduced) each year, so clearly the figures could be higher.

Twenty-two of the study plants were required to report to TRI; the 16 that reported the annual amount of waste reduced to INFORM were among them. These 16 study plants reported a total of 24.6 million pounds of TRI chemicals released in or transferred as waste for the year 1988.

To compare the TRI figures with the reduction in waste reported to INFORM, several points should be noted. First, TRI releases and transfers refer to amounts of the chemicals after any on-site waste treatment. The amount of waste actually generated before on-site treatment can be as much as one-third higher. This figure is based on the fact that for all facilities submitting TRI reports to EPA in 1988 with information on their source reduction and recycling activities (as well as on releases and transfers), releases and transfers were 66% of their reported waste generated.[2] For the 16 INFORM study plants, then, it is estimated that, if 24.6 million pounds of releases and transfers were reported, then 37.3 million pounds of TRI chemicals were generated as waste.

Second, the 128.7 million pounds of waste reduced per year reported to INFORM by the study plants represent the total volume of waste (including inert materials such as water and soil), not just the amount of the chemical constituents in the waste, as is contained in the TRI reports. From the data received from the study plants, the 128.7 million pounds can be broken down as follows.

- Almost 3.9 million pounds are reported as the actual chemical constituents in the wastes that were eliminated as a result of source reduction activities.

---

2 Environmental Protection Agency, *Toxics in the Community,* September, 1990, p. 306.

- About 84.0 million pounds were RCRA wastes, as defined by the US EPA, which include the amounts of inert materials such as water and soil as well as the chemical constituents. Taking 10 percent of the total volume as a reasonable estimate of the amount of actual chemical constituents in these wastes yields a figure of 8.4 million pounds of industrial chemicals in this waste.
- Another 30.4 million pounds of waste were reported to have been avoided entirely (that is, these wastes were never generated in the first place) through process design prior to full-scale construction and operation.
- An additional 10.4 million pounds is the amount of inert material in a particular wastestream at the Dow plant (separation of amount of inert material from amount of chemical constituents based on calculation).

Thus, only the subtotal of 12.3 million pounds (3.9 million pounds of chemical constituents and 8.4 million pounds of chemical constituents in RCRA wastes) can be compared to the TRI waste generation numbers.

Adding the 12.3 million pounds of toxic chemical waste reduced to the estimated 37.3 million pounds of TRI waste, the amount of waste that would have been generated with no source reduction activities in place is estimated as 49.6 million pounds. If the ratio of releases and transfers to waste generated of 66 percent is applied, then the estimated releases and transfers would have been 32.7 million pounds instead of the 24.6 million pounds reported to TRI. This is one-third (33 percent) higher than if the source reduction activities had not taken place. (Appendix B details the calculations used for this analysis.)

While 33 percent less toxic chemical waste entering our air, land, and water from these plants is an important improvement, data from the annual reports on chemical industry waste generation compiled by the Chemical Manufacturers Association (CMA) and from RCRA hazardous waste generation reports indicate that the overall amount of industrial hazardous and toxic waste generation in the United States continues to climb. Further promotion and adoption of source reduction programs and strategies could play a role in reversing this trend and reducing the burden of toxic chemicals on our environment.

Conclusions

Overall, the organic chemical plants INFORM studied have dramatically decreased their releases of toxic and hazardous wastes, saved money, and increased product yields through source reduction. Yet these plants represent only a fraction of the waste-generating facilities in the United States and, indeed, around the world. Based on the growing concerns about these wastes, and on the specific findings detailed above, INFORM draws several conclusions.

1. Source reduction offers waste-generating facilities continuing opportunities to significantly reduce the amounts of toxic and hazardous wastes they generate and subsequently release into the air, land, or water, or transfer to treatment or disposal facilities. Large reductions continue to be achieved even at plants that have been implementing source reduction for many years.

2. Source reduction also offers waste-generating facilities continuing opportunities to rapidly reduce costs and increase production efficiency.
    - The cost of implementing many source reduction activities is low (no capital investment was required for one-quarter of the source reduction activities in INFORM's study for which capital cost data were provided, and investments of less than $100,000 were required for just under half of the source reduction activities).
    - Companies can recoup their initial investments rapidly (in 6 months or less for two-thirds of the source reduction activities in INFORM's study for which information on payback periods was reported).

- Product yields increase (for 97 percent of the source reduction activities in INFORM's study for which information on changes in product yield was provided).
- Implementation times, including research and development, are short (6 months or less for nearly two-thirds of the source reduction activities for which plants reported implementation time information to INFORM, and 6 months to 3 years for another 30 percent).

3. Waste-generating facilities are likely to find more source reduction opportunities if they establish source reduction programs with several key features: full cost accounting systems, employee training and incentive programs, and high-level leadership that includes both operations and environmental managers. Plants in INFORM's study that adopted any of these program features reported statistically significantly more source reduction activities than plants lacking them. Plants that adopted five or more of the eight program features tracked by INFORM reported, on average, 2.5 times as many source reduction activities as plants with fewer program features.

4. Enforcement of environmental regulations governing the treatment and disposal of wastes is one of the key steps government can take to promote source reduction. Such regulations have served as an important factor motivating company officials to look for ways to reduce waste: rather than standing in its way, such regulations were the fastest growing incentive for implementing source reduction activities at INFORM's study plants between about 1978 and 1990.

   In particular, environmental regulations limiting the inexpensive disposal of hazardous wastes tend to encourage source reduction; thus, some regulations, such as those that permit burning wastes as fuel and deep well injection, discourage source reduction. It is likely that even more source reduction might be expected at these and other plants if such waste management options were less accessible than they are now.

   INFORM's 1987 report, *Promoting Hazardous Waste Reduction: Six Steps States Can Take,* identified other strategies governments can use, in addition to regulation, to encourage waste-generating facilities to establish source reduction programs with the features discussed here.

5. Managers of waste-generating facilities can continually find many low-technology, efficiency-oriented source reduction opportunities when they look for them; process changes, operations changes, and equipment changes accounted for 87 percent of the source reduction activities reported to INFORM. Such efficiency-oriented opportunities continued to be found by plants that had previously achieved significant reductions in waste generation.

6. Additional research is needed to understand the full potential of the more innovative source reduction techniques: product changes and chemical substitutions. A thorough analysis of the benefits, motivating factors, and obstacles associated with these techniques was not possible since only 13 percent of the source reduction activities in INFORM's study involved such changes. It is worth noting, however, that all of the product changes reported to INFORM with percent reduction data achieved complete elimination of the target wastestream.

PART II PLANT PROFILES

Introduction to the Plant Profiles

Part II contains information on each of the 29 chemical plants studied by INFORM. For plants that cooperated with INFORM's study, the information was collected during interviews at the plants; no additional data were obtained from outside sources. For plants that did not grant an interview, however, written materials from government sources and from the plant itself, when available, were used. In particular, for those plants that did not cooperate, the profiles include data on environmental releases and transfers of toxic chemicals reported to the US Environmental Protection Agency in the Toxics Release Inventory for 1987 and 1988. Each profile text indicates the degree to which the plant cooperated with INFORM's study.

The initial interviews took place during 1987 and 1988. Each plant was then sent a copy of its profile for review, comments, and correction. In 1990, the revised profiles were sent to the plants with a request for any updated information on new source reduction activity. Except for Exxon, all of the plants that cooperated with INFORM during the initial review stage responded to this final request for information.

References to products and processes in the profiles are intentionally general in some places because of the companies' concerns about proprietary information. The profiles are not meant to provide an assessment of engineering technology. Rather, they focus on the reasons particular source reduction activities were undertaken from an institutional and organizational standpoint, the amounts of waste reduced, the costs and savings associated with these source reduction activities, and the factors influencing and inhibiting the implementation of source reduction activities at individual plants.

Organization of the Profiles

The organization of each profile reflects this report's broadened focus compared to INFORM's 1985 *Cutting Chemical Wastes:* including the institutional framework of policies and programs that encourage source reduction within a plant as well as individual source reduction activities and their motivation.

Each profile begins with a summary, a quick look at some of the key facts about the plant, followed by a discussion of the plant's products and operations. The next three sections involve the plant's source reduction program features. The first looks at the company's environmental policy, including corporate as well as plant-specific policies, where applicable. The next section describes the plant's system of materials data collection, including information on how it keeps track of materials used and waste generated. The third section covers other source reduction program features such as leadership and worker involvement.

The profiles then turn to each plant's reported source reduction activities, with a table summarizing key information about each of the activities and text discussion. Only the activities newly reported to INFORM for this study are listed in the tables and discussed; information on the source reduction activities described in *Cutting Chemical Wastes* is not repeated here. Each table is set up to include a brief description of each source reduction activity, along with the type of waste affected, the source reduction technique used, the specific waste reduced, the amount of waste reduced, the change in product yield, the capital cost, the annual savings, the motivation for undertaking the activity, and the time needed for implementation (including research and development). The plants did not provide information on each of these topics for each source reduction activity; blanks in the tables indicate this absence of data. Information from government reports, where available, was used for plants that did not cooperate with INFORM.

The next section of the profiles includes information on waste management activities, other than source reduction, that the plants reported to INFORM. Finally, the last three sections contain comments by plant officials on technical assistance used or provided by the plant, on governmental activities affecting the plant's source reduction activities, and on the outlook for further source reduction efforts in the near future. These comments reflect only the views of the individuals interviewed; they have not been reviewed by others outside the corporation or analyzed by INFORM.

All the information contained in the profiles was provided by the plants to INFORM or to governmental agencies and has not been verified through other sources.

Plant Information Contained in INFORM's Earlier Study

As mentioned above, these profiles only include information on the 137 source reduction activities that the plants described for the first time for this report. Forty-four other source reduction activities were described in *Cutting Chemical Wastes* and form part of the analysis in Part I. For those readers wishing to refer to that earlier INFORM report, the list below indicates any differences in plant names (due to new ownership or other reasons) between the two books.

Environmental Dividends	*Cutting Chemical Wastes*
American Cyanamid Company	American Cyanamid Company
Aristech	USS Chemicals
Atlantic Industries	Atlantic Industries
Bonneau Dye Corporation	Bonneau Dye Corporation
Borden Chemical Company	Borden Chemical Company
Chevron Chemical Company	Chevron Chemical Company
Ciba-Geigy Corporation	Ciba-Geigy Corporation
Colloids of California	Colloids of California
Def-Tec Corporation	Smith and Wesson Chemical Company, Inc.
Dow Chemical USA	Dow Chemical USA
E. I. Du Pont de Nemours and Company	E. I. Du Pont de Nemours and Company
Exxon Chemical Americas	Exxon Chemical Americas
Fibrec, Inc.	Fibrec, Inc.
Fisher Scientific Company	Fisher Scientific Company
Frank Enterprises, Inc.	Frank Enterprises, Inc.
Hart Chem/J. E. Halma	J. E. Halma Company, Inc.
ICI Americas, Inc.	Stauffer Chemical Company
ICI Resins US	Polyvinyl Chemical Industries
International Flavors and Fragrances, Inc.	International Flavors and Fragrances, Inc.
Max Marx Color and Chemical Company	Max Marx Color and Chemical Company
Merck and Company, Inc.	Merck and Company, Inc.
Monsanto Company	Monsanto Company
Morton International, Inc.	Carstab Division
Perstorp Polyols, Inc.	Perstorp Polyols, Inc.
PMC Specialities Group	Sherwin-Williams Company

Environmental Dividends

Rhône-Poulenc, Inc.
Scher Chemicals, Inc.
Shell Chemical Company
Unocal Chemicals

Cutting Chemical Wastes

Rhône-Poulenc, Inc.
Scher Chemicals, Inc.
Shell Chemical Company
Union Chemicals Division

AMERICAN CYANAMID COMPANY
Marietta, Ohio

Summary

American Cyanamid's Marietta plant, built in 1915, is located in the town of Marietta in southeastern Ohio near the Ohio River and the border with West Virginia. The complex of manufacturing facilities at the site produces vulcanized vegetable oils, ultraviolet absorbers, and organic intermediates. The plant increased production from 1983 to 1987, but in more recent years has sold off most of the Formica brand of products and all of its water-soluble organic dyes product line, with a related decrease in employees to 90.

American Cyanamid's environmental policy is driven by the need to comply with governmental regulations. Because RCRA forced attention on their disposal of solid waste in landfills, company officials have developed materials tracking systems and cost accounting procedures for solid waste. More recently, the plant's wastewaters have come under scrutiny by the municipal sewage treatment plant and these are being incorporated into their cost accounting system. In seeking the most cost-effective way to change their pattern of waste disposal, plant officials investigated alternatives such as treatment or incineration but found several source reduction methods not only to be cost-effective but to result in actual cost savings.

Overall, solid waste generation has gone from 952,000 pounds in 1983 to 762,800 pounds in 1987, a decrease of 20 percent. During this same time period, production levels have gone from 5.5 million pounds in 1983 to 10 million pounds in 1987, an 82 percent increase.

This plant was the first of American Cyanamid's plants to implement a corporate materials-tracking system. Waste audits are conducted on-site, using plant personnel for the most part. Generally, its employees as well as corporate staff offer technical assistance related to source reduction to others both within and outside the corporation.

American Cyanamid's Marietta plant cooperated in this INFORM research for the first time, having not granted an interview in INFORM's original study. The company conducted a tour for INFORM researchers of its Marietta, Ohio facility, and commented on corporate-wide programs as well as those at the Marietta plant. Plant officials reported a total of six source reduction activities reducing 803,600 pounds of waste, saving the company $220,000 each year.

Products and Operations

American Cyanamid's Marietta plant is one of 40 domestic plants in the company's Chemicals Group. This group is, in turn, the largest of four groups in American Cyanamid; the others are medical, agricultural, and consumer products. The Marietta plant manufactures vulcanized vegetable oils, ultraviolet absorbers, and organic intermediates. The volume of individual chemicals used ranges from 1,000 pounds to 1-2 million pounds per year. From 1983 to 1987, the volume of production almost doubled, rising from 5.5 million pounds to 10 million pounds per year. The number of employees at the plant increased by 17 percent, from 128 in 1983 to 150 in 1987. However, in 1990, water-soluble organic dyes, a major product line, were sold, with a related decrease in employees to 90.

Environmental Policy

As a member of the Chemical Manufacturers Association, American Cyanamid has adopted the "Responsible Care" program which calls for companies to set and pursue multimedia goals for reducing waste. The Marietta plant engineer reported to INFORM that no plant-wide source reduction goals have been set at Marietta, although, in recent years, higher levels of management, like himself, have become more active in this area on a day-to-day basis.

American Cyanamid's management policy relating to environmental goals, written in 1977, is to manage wastes at its plants in accordance with all regulatory requirements and to manage them in the most cost-effective manner. This policy has been updated frequently since then as regulations have changed and the company has become increasingly aware of environmental concerns. The primary concern in 1977 was solid waste going to landfills, but the updates have expanded the focus to include wastewaters and air emissions to ensure that they comply with state and federal regulations, as they were developed. Based on economics and regulation, the corporate goals are to minimize wastes (through reduction and treatment) and to render hazardous wastes "nonhazardous" according to RCRA definitions so that they no longer fall under RCRA regulations. For American Cyanamid, regulations have clearly forced the issue in many cases but, even without specific regulations, costs have been driving a search for higher yields and reduced wastes.

While the plant has no overall source reduction goals, it has source reduction targets for each project. The current project at the Marietta plant is to eliminate its largest air release.

Materials Data Collection

American Cyanamid is developing a corporate-wide database that will contain information tracking waste generated at every plant on a process basis. The Chemicals Group was chosen as the first corporate area for attention because it is the largest group in the corporation, and the Marietta plant was the first to implement this computer materials tracking system. It has since been applied to all 40 plants in the Chemicals Group. American Cyanamid is also developing the computer tracking system for its other three groups.

Installation of the waste-tracking system at Marietta began in 1983 and was completed in 1986 at a cost to the corporation of $0.5 million. So far, the system has only been used for RCRA-regulated solid and semiliquid waste disposed of off-site, but it has the capability to add tracking of air and water wastes at a later date. The system tracks, on a daily basis, inventory, individual wastes, hazardous solid waste manifests, the return of the manifests, and verification of the waste disposal site and costs. Maintenance operations, as well as other nonproduction operations such as loading/unloading, are included in the tracking system. Tracking is done by several categories: process, waste name, waste category, and individual constituents for each waste.

The waste-tracking system has given the plant the ability to do full cost accounting for all waste, although such accounting is applied only to the solid and semiliquid RCRA waste at present. For this waste, costs are allocated far more specifically than in the past, no longer just to the level of a building on the site but to the process operation level. On-site treatment costs are part of the allocation, but regulatory compliance, insurance, costs of accidents, waste clean-up, and public or customer relations dealing with waste issues are not specifically allocated to processes. Product managers receive reports stating the full costs of their operations in terms of dollars per pound of product; in some cases, the plant official noted, they have been surprised by the full cost of their product. The cost accounting system also allocates wastewater costs to processes, again in terms of dollars per pound of product, but not as accurately as solid wastes since the tracking system does not include wastewater. Costs of air emissions are not allocated to processes.

American Cyanamid has found that when the specific products produced change daily, as they do at the Marietta plant, it takes the company up to 3 or 4 years to develop a database sufficient to identify possibly beneficial source reduction actions.

Other Source Reduction Program Features

In addition to having a written source reduction policy, materials accounting (but not materials balance), and cost accounting, American Cyanamid has fully or partially adopted all of the other four source reduction program features tracked by INFORM: leadership, employee involvement, environmental goals, and an environmental program.

American Cyanamid has a corporate environmental section, as part of the Safety, Health and Environmental Department of the Chemicals Group. This section reports to a vice-president, and consists of 18 people. One-third have environmental degrees and the rest have technical degrees. It develops environmental management policies and oversees the corporate-wide tracking system as well as performing audits and giving advice to plants on dealings with governmental agencies. It also disseminates progress reports and case studies of source reduction to all the plants in the Chemicals Group.

The Marietta plant has had a hazardous waste handling program since it closed its on-site landfill in 1979 because it could not meet impending state regulations. This program involved sending RCRA-regulated wastes off-site to secure landfills and inspecting conditions at the landfills to ensure that the wastes were being disposed of according to American Cyanamid standards. The program included annual training classes for plant employees for the proper handling of waste.

At the time of the program's inception, the primary concern was to find out what waste was being generated and to make sure that wastes were going to suitable disposal sites. However, during the 1980s, the plant faced regulatory changes, increased scrutiny, and increasing disposal costs. In addition, hazardous wastes are now much more strictly defined under government regulations. Because of these developments, American Cyanamid instituted a new program in 1986 to review all plant operating procedures. The reviews, conducted by the technical and production departments, include all waste generation and handling; results are reported to the plant manager. Responsibility for implementing any recommendations made in the reviews lies with the production supervisor.

The waste review process at American Cyanamid focuses on both new and existing products and processes. Multimedia hazard reviews are conducted for all new products and the processes that may be used to make them. These reviews use a fault tree analysis technique developed only fairly recently. The analysis asks "what if" for many different scenarios of operations, marketing, and environmental releases. Potential environmental problems are examined in light of environmental regulations, possible useful by-products, and alternate disposal methods.

For existing processes, this type of review is conducted every 3 years. Any new capital expenditure program also comes under such a review and requires an accompanying letter from the corporate environmental group. The letter reviews waste generation and disposal problems and is sent to the corporate managers who must approve such expenditure programs.

A four-member "waste minimization" committee was established in 1984 in response to rising waste disposal costs and problems. The four members have backgrounds in chemistry, public health, production operations, and engineering. The committee conducts general waste handling discussions with plant employees, brainstorming, and specific project follow-ups. The results are reported to the plant manager. The responsibility for carrying out the recommendations of the committee lies with the environmental services manager.

Source Reduction Activities

American Cyanamid reported implementing six source reduction activities to INFORM. They are summarized in Table II-1 and described below. No source reduction activities were reported to INFORM for the 1985 report.

American Cyanamid reformulated two yellow dyes manufactured at the Marietta plant when the city of Marietta installed secondary treatment at its sewage treatment plant and prohibited the plant from sending certain of its wastewater streams to the plant. Since the substances used to make the dyes were listed as "hazardous" under RCRA regulations because of corrosiveness or ignitability (depending on product form), American Cyanamid concluded that treatment of the yellow dye waste and subsequent disposal off-site would not be cost-effective. In 1987, the company identified yellow dyes with a different chemical makeup that could be produced using the same equipment and that would involve generation of a more acceptable effluent. The dye reformulations reduced the amount of waste in the form of waste cake by about 500,000 pounds per year and the amount of ammonia in wastewater effluent by 100,000 pounds per year.

Regulations and the cost of disposal options motivated the company to eliminate use of nitrobenzene as a solvent for another yellow dye produced at the plant. Nitrobenzene waste had always presented a disposal problem. Originally, it was put in on-site impoundments. Then, under the 1976 RCRA regulations, any site storing "hazardous" chemicals, which included nitrobenzene, had to be classified as a hazardous waste site. This made the plant incur related expensive handling and monitoring costs. When nitrobenzene waste was slated to be banned from land disposal altogether by 1990, the plant turned to even more costly incineration. But even this was proving inadequate. The nitrobenzene waste was produced faster than it could be recovered or disposed of and storage capacity was limited. With the new process adopted in 1987, nitrobenzene was replaced by an organic solvent, acetic acid, which is reused in the same process.

This source reduction measure required several months of research and development work in American Cyanamid's laboratory and entailed $100,000 in capital equipment costs to implement. However, it reduced the cost of production of this yellow dye significantly. The company now saves $200,000 per year, mainly from reduced disposal costs but also from avoiding energy costs that were necessary to recover nitrobenzene. The old process was also the largest single contributor of BOD (biological oxygen demand) in American Cyanamid's wastewater discharge to the city of Marietta's sewage treatment plant. The BOD loading from this yellow dye was reduced by 90 percent, or 120,000 pounds per year, through this source reduction measure.

Spill control measures have also reduced hazardous wastes at this plant. Prior to 1988, the average amount of materials lost as waste annually due to spills in the tank area was about 1,000 pounds. In 1988, the tank area was compartmentalized and contained. The tanks were put on pads that catch spills, and any spills are pumped to a sump for recovery and recycling when possible. The company was motivated to add the tank farm pads so that its new wastewater treatment system would not have to treat spilled wastes and so that spills would not go onto the ground. As part of the same spill control project, run-off from the tank area that used to flow into ponds on-site and then to the municipal sewage treatment plant has been isolated in order to reduce the volume and cost of effluent treatment. The containment has reduced wastes by about 90 percent or 900 pounds per year. The construction costs for this project were $700,000 and the estimated savings are $10,000 per year.

American Cyanamid accomplished a different type of waste reduction — a reduction of hazardous wastes generated by its customers — by modifying a product. In one of its products (the company would not specify which), the use of cellosolve acetate as a solvent was replaced by use of the nonhazardous solvent ethylene glycoldiacetate. This source reduction activity was motivated by a desire to increase product marketability. American Cyanamid no longer has to label the product as hazardous, and customers no longer have to dispose of any leftover product as a hazardous waste. While costs savings for customers are unknown, American Cyanamid saves $10,000 per year as a result of this change.

One source reduction project affected air emissions, although the material reduced was not toxic. Since 1980, the Marietta plant changed from producing dry to producing mostly wet, nontoxic products, thereby reducing dust emissions. The research to accomplish this for one major product line took 2 years. The primary reason for the changes has been customers' preferences, but air emissions of dusts have been reduced as a consequence, by about 90 percent, or 2,000 pounds per year.

The sixth source reduction activity, a simple operating change instituted in 1985, has led to a 90 percent reduction in liquid hazardous wastes from American Cyanamid's laboratory. Prior to 1985, the samples used in the laboratory were thrown away or washed into the sewer. In order to reduce the toxicity of the lab's wastewater to meet the new public sewage treatment plant requirements, samples are now reintroduced back into the process to the extent feasible. Those that cannot be reused are disposed of in lab packs (special containers for laboratory waste) as a hazardous solid waste. Plant officials find it easier to control and know what happens to wastes handled in this way than when they were disposed of in wastewater; the lab packs amount to no more than one drum per year.

Table II-1 American Cyanamid (Marietta, OH): Source Reduction Activities

Waste Medium (SR Type) Year	Source Reduction Activity	Specific Waste Reduced (Hazardous or Nonhazardous)	Percent Waste Reduced	Amount Waste Reduced
Solid (PR) 1987	Replaced two yellow dyes with another yellow dye using the same equipment.	Six waste cakes (N) Ammonia (H)	100%	500,000 lb/yr 100,000 lb/yr
Solid/water (PS,CH) 1987	Modified a yellow dye manufacturing process to substitute new solvent for nitrobenzene. New solvent recovered through in-process recycling.	Nitrobenzene waste (H) Biological oxygen demand (BOD) (N)	100% 90%	200,000 lb/yr 120,000 lb/yr
Water (OP) 1988	Compartmentalized tank farms onto pads to contain spills, which are recovered for use when possible.	Spilled organic and inorganic materials (H)	90%	900 lb/yr
Water (CH) 1985	Replaced a hazardous solvent, cellosolve acetate, with a non-hazardous solvent, ethylene glycoldiacetate.	Cellosolve acetate (H)	Unknown	
Air (PS, PR) 1980+	Switched from producing dry products to producing mostly wet products.	Dust (N)	90%	2,000 lb/yr
Water (OP) 1985	Recycle quality control samples from lab back to process wherever possible.	Quality control samples (H)	90%	9 drums/yr (2,700 lb/yr)

Key to source reduction types: CH, chemical substitution; EQ, equipment change; OP, operational change; PR, product change; PS, process change.
A blank indicates that the plant did not provide information.

Other Waste Management Practices

In addition to its source reduction activities, American Cyanamid also reported the following waste management practices.

One of the early results of the waste review process at this plant involved the handling of solid wastes. Procedures were established to ensure that the content of all containers is known and that all wastes classified as hazardous are put in the right containers without contaminating other wastes. As an operational procedure, the categories of wastes at the plant were changed from three (general, general contaminated with color, general contaminated with hazardous materials) to two (general and hazardous). Also, a continuing training program was designed to ensure that employees follow the correct procedures. The first-year cost savings as a result of these new procedures amounted to $40,000.

High disposal costs motivated two recycling projects. In one, equipment to precipitate and isolate zinc generated as waste in a clarification process to produce brighteners was installed at a cost of $50,000 to $60,000. American Cyanamid expects to sell the zinc to a recycler. While

Change in Yield	Dollars Saved	Dollars Spent	Motivation	Comments	Time Needed for Implementation
			Can no longer send certain wastes to the public sewage treatment plant.	Treatment and disposal not cost-effective.	6 mo
None	$200,000/yr	$100,000 for equipment	Waste cake banned from landfilling, costs and problems of incineration. New limitations of BOD at city sewage treatment plant.	Cost savings from reduced disposal costs and energy savings from not having to recover nitrobenzene.	Several months
	$10,000/yr	$700,000	Project to avoid treatment of on-ground spills in new on-site wastewater treatment plant.		6 mo for design
	Unknown for customers, $10,000 at American Cyanamid	$0	Increases marketability of product by eliminating hazardous waste label.	Change eliminates need for customer to dispose of any leftover product as hazardous waste.	1 yr
Unknown	Unknown	$0	Customers preferred wet products.		2 yr for one major product line
	Unknown	$0	Reduce toxicity and volume of wastewater stream to meet new requirements of the city's sewage treatment plant.		None

the cost of isolating the zinc is greater than the revenue received from the recycler, the net cost is still less than disposal costs would be.

The second recycling project involves recovering about 1 million pounds per year of aluminum hydroxide, used as a raw material for cement manufacture, by diverting a wastewater stream. This project was undertaken when the Marietta sewage treatment plant imposed a limit on the amount of this chemical it would accept, and the company found the cost of landfill disposal to be unacceptably high. Turning to a recycling solution, American Cyanamid implemented the project within 3 months in 1989. Equipment costs were just $2,000, while savings amounted to over $100,000 per year in landfilling expenses.

Technical Assistance

American Cyanamid provides other companies with technical information related to source reduction. It has attended and given reports on its program at the annual League of Women

Voters conferences held at Woods Hole, Massachusetts. It also holds conferences for representatives from its plants nationwide to exchange ideas. One of the direct responsibilities of the environmental compliance manager on the corporate staff is the transfer of technology within the corporation.

American Cyanamid also reports that it has found useful and has been active in the conferences on technical issues that have been held by the Ohio Manufacturers Association and other trade associations and professional societies.

Company Comments on State and Federal Regulations

American Cyanamid described the process of obtaining permits from the state of Ohio as a very slow process and said that anything that would ease the burden of moving a project through the state and federal agencies would be welcome. The problem, from American Cyanamid's perspective, is that technical issues need technical answers and that the high rate of turnover in state agencies means that high-level people who are not familiar with a project have to give answers. State personnel at lower levels, who might be familiar with a project, cannot make decisions. The company also noted that because the state of Ohio was not delegated with the responsibility for water pollution control programs by the federal EPA until 1989, US EPA Region V personnel used to start asking questions once the questions of the state personnel were answered. This was a problem with American Cyanamid's attempt to install a new wastewater treatment system at the Marietta plant. While there is a rule that the state must reply to the plant's documents within 90 days, plant officials report that the permit application process took over 2 years.

Future

American Cyanamid expects further environmental legislation and is trying to anticipate the changes. To do this, the corporate staff keeps close contacts with trade associations and lobbyists at both the federal and state levels. In particular, air emissions are seen as an area worthy of study. American Cyanamid has undertaken a special study of volatile organic compound (VOC) emissions at the Marietta plant and the other plants in the Chemicals Group. Currently, the Marietta plant has VOC scrubbers in a closed system, and the plant is annually inspected by state personnel for any changes in operations affecting air emissions.

American Cyanamid sees source reduction at its plants as directed by the business climate. As its competitors reduce production costs, it will have to find ways to do so too, and it has found an opportunity for reducing costs in reducing wastes. This is especially true for solid wastes because costs to landfill even nonhazardous solid wastes are rising. American Cyanamid also anticipates a new emphasis on hazardous chemicals in wastewater as pretreatment limits are set and organics in wastewater are better defined.

ARISTECH
(formerly USS Chemicals)
Haverhill, Ohio

Summary

Aristech's Haverhill (Ohio) plant, built in 1962, is located in south central Ohio. The plant is situated on the Ohio river and gets most of its raw material (cumene) from river barges while it ships its products out on trucks. This plant produces phenol, aniline, acetone, and bisphenol-A (BPA), and its secondary products are alpha-methylstyrene (AMS) and diphenylamine (DPA). While these products have not changed, overall production increased between 1982 and 1987: from 10 percent for acetone to 200 percent for DPA. Employment increased 15 percent in the same time period.

The plant became part of Aristech when the USX Corporation divested itself of its chemical operations. Aristech, as a company, does not have a written environmental or source reduction policy. Instead, its focus on waste tracking and monitoring springs from a quality improvement program tied to its business policy of economic efficiency through maximizing the conversion of feedstocks or raw materials into saleable product. This quality improvement program is designed to give employees tools to monitor progress in product output (and hence, reduced waste), and encourages participation by all of its employees in all aspects of plant operations.

Aristech reported a total of ten source reduction activities since the publication of INFORM's *Cutting Chemical Wastes*, resulting in an annual reduction of waste by 23.8 million pounds, saving the company $3.5 million annually. USS Chemicals, the plant's former owner, had reported six source reduction activities for the earlier study, reducing 3.1 million pounds of waste and saving $278,750 annually.

Products and Operations

In December, 1986, the USS Chemicals Division of USX (formerly U.S. Steel), as a result of divestiture, became an independent company named Aristech. Aristech has ten plants in eight states. The Haverhill plant is the nation's largest producer of phenol, accounting for more than 15 percent of US industrial capacity. Phenol is widely used to manufacture synthetic resins and industrial chemicals. In 1987, the phenol capacity at the plant was expanded and is now 630 million pounds. The plant also produces acetone, a coproduct of phenol in the cumene oxidation process, in a quantity totaling more than 10 percent of US industry capacity. Acetone is used as a solvent in paints, pharmaceuticals, and nail polish remover. Another coproduct, alpha-methylstyrene, used to manufacture industrial paints and resins, amounts to more than 45 percent of US capacity.

Cumene hydroperoxide (CHP) is created at the plant as an intermediate product of phenol and acetone. In 1987, the Haverhill plant installed equipment to increase the purity of the CHP to a high enough grade to be able to sell it. The Haverhill plant also produces bisphenol-A (BPA), a component in the manufacture of industrial paints and resins, accounting for more than 10 percent of US industry capacity; the process uses phenol and acetone as feedstocks. A two-stage BPA expansion program was completed in 1989.

The company's aniline unit started up in 1982, and diphenylamine (DPA) was produced in the aniline unit beginning in 1983 (DPA is used to process synthetic rubber and foam).

The type of products produced at Haverhill has not changed since INFORM's first interview for the 1985 study, but production levels have increased. In 1987, Aristech's Haverhill plant employed 257 people, an increase of 15 percent since 1983. The company's sales in 1987 were $918.8 million, an increase of 28 percent since 1983. Annual phenol production has increased about 15 percent to 630 million pounds annually. Acetone production has increased about 10 percent to 394 million pounds annually; production of BPA has increased 79 percent to 215 million pounds. Production of DPA has increased by 200 percent to 4 to 5 million pounds annually, and the amount of cumene used as a raw material has increased by 30 percent to 917 million pounds. Aniline production remains at 200 million pounds per year.

Environmental Policy

Aristech does not have a written corporate policy favoring reduction of waste at source over other waste management strategies — in fact, the company has no written environmental policy at all. The officials interviewed by INFORM stated, however, that its policy is to be in full compliance with environmental laws and to integrate environmental concepts into all business practices. Aristech expects that each plant will ensure such compliance and corporate staff provide resources and assistance needed by the plants. Haverhill's plant manager pointed out that source reduction pursuits fit neatly into the company's business philosophy, which places first priority on converting feedstock to product, second priority on finding product uses for wastes, and third on reusing and recycling as much of the by-products as possible.

Aristech has a corporate-level office of health, safety, and environmental affairs, formed in June, 1987, 6 months after Aristech became an independent company. The vice-president in charge of this office reports directly to the chief operating officer. In addition, the Haverhill plant has an environmental group, formed in 1985 because of the increasing work involved in RCRA compliance. It has five full-time employees. The approach of the environmental group has been to become specialized in each regulated area (air, water, and RCRA) in order to comply with the laws.

Aristech officials reported that the formation of the separate company, following divestiture from USX, has helped environmental efforts at the Haverhill plant in two ways. First, this smaller, leaner firm can make decisions faster and, second, the formation of the corporate office of environmental affairs let plant personnel know that the corporation will support their efforts.

Under USX, the plant had to depend on corporate attorneys for legal compliance. USX, being primarily a steel company with fairly fixed production processes, had a top-down "command and control" approach that did not adapt itself well to the entirely different world of chemical manufacture. While the Haverhill plant, under USX ownership, never had a problem in getting the resources to make necessary changes, its managers find that Aristech, as a company entirely devoted to chemical manufacture, can respond more quickly to changes in demand and that knowledge of ways to reduce waste in chemical processes is available through a specialized environmental staff at the headquarters level.

Materials Data Collection

Aristech established a materials accounting system in 1987. The accounting system begins with a yearly environmental audit conducted at the plant. The audit's primary focus is to identify areas where system improvements can be made in all phases of operations, including reducing air emissions and wastewater discharges as well as solid waste. The audit teams consist of five or more outside consultants, each specializing in an environmental area (RCRA; Toxic Substances Control Act, or TSCA; air; National Pollutant Discharge Elimination System permits, or NPDES; and groundwater) and an Aristech engineer. The teams review compliance records and conduct site visits.

As part of Aristech's business policy, the audit team looks at waste for the potential of economic recovery by maximizing the amount of feedstock going into the product. Then product uses for the waste are explored, and only then are recycling and reuse considered. After the final report is distributed, the plant (with corporate help) prioritizes work items and submits progress reports on a monthly basis. The review report lists each source reduction project with a priority number, and monthly progress reports are submitted to headquarters.

In addition to the yearly audit, the Haverhill plant's environmental group conducts its own inspections of plant operations on a routine basis, as often as three times a week.

A special "waste minimization study," part of an overall corporate directive, was begun at Haverhill in 1987. The study's first phase (characterizing processes and identifying constituents in wastestreams at the plant) has been finished and the second phase (developing specific source reduction projects) has begun. Outside consultants are being used for this study because, as a small company, Aristech does not have the resources to devote to such an effort. Also, outside consultants can take a fresh look at problems. However, the study takes longer than it would with Aristech staff because outside consultants have to be educated about the specific processes and procedures at the plant.

The Haverhill plant also has a cost accounting system that covers all wastes except those released to the air. Waste treatment and disposal, including permit fees, are a separate budget item; capital expenditures are separated from routine costs. All changes in costs are reviewed at monthly meetings, and every manifest form for off-site shipment of RCRA wastes must be signed by the plant manager.

Other Source Reduction Program Features

Of the remaining key source reduction program features outlined by INFORM, Aristech has fully or partially adopted leadership and employee involvement programs, but does not set numerical goals for source reduction.

An innovation at Haverhill, established under Aristech ownership, is the Aristech Total Performance (ATP) program. ATP takes a systems approach to defining quality in every aspect of the business based on statistical process controls (SPC) and other methods. The SPC techniques monitor production with statistical tracking and analysis so that operations personnel can identify areas of needed improvement and eliminate variations in processes that would otherwise result in unacceptable product quality. SPC was first used on a pilot basis in 1984 at another Aristech plant where it was successful even though SPC applicability to batch processing had been doubted. SPC was introduced in the Haverhill plant at the beginning of 1987. The ATP program also involves training each employee in both attitude and methods, including SPC techniques.

Many process and other changes that have accomplished source reduction have resulted from this new ATP approach. The plant manager reported that the largest gains in source reduction in the future at the Haverhill plant are expected to be the result of the "waste minimization study," which is one aspect of ATP. The plant's environmental compliance program and its ATP program focus on wastes as costs, as well as on unacceptable product quality, as a way of helping identify ways to reduce wastes.

Aristech also has an employee suggestion program that has proven over many years to be a rich source of ideas for reducing wastes. This program, which is now incorporated into the ATP program, began as a corporate program under U.S. Steel management and was called Suggestions for Cost Reduction (SCORE). It provides monetary rewards to nonmanagement employees for suggestions on ways to save the company money. The plant manager stated that this program has been particularly successful at Haverhill because employee turnover at this plant is low, since it is in a high unemployment area, and employees become well trained in the plant operations.

The "waste minimization study" has also been a factor in promoting waste reduction since employees help characterize each wastestream. They then tend to begin looking with new interest and awareness for ways to reduce wastestreams. In one instance reported to INFORM by the plant manager, an employee suggestion was first turned down, but the employee was persistent and eventually management implemented the suggestion, which resulted in reductions of more than 1.5 million gallons of hazardous wastewater a year.

Source Reduction Activities

Since 1984, when six source reduction activities were reported, Aristech reported ten source reduction activities (summarized in Table II-2 and detailed below) in three separate areas: heavy ends wastes, wastewaters going to injection wells, and off-site waste disposal. These have produced major gains: waste generation at the Aristech Haverhill plant declined approximately

Table II-2 Aristech (Haverhill, OH): Source Reduction Activities

Waste Medium (SR Type) Year	Source Reduction Activity	Specific Waste Reduced (Hazardous or Nonhazardous)	Percent Waste Reduced	Amount Waste Reduced
Solid (PS) 1984	Recycle bisphenol-A (BPA) mother liquor before last step in the process.	Heavy ends, organic waste (H, N)		
Solid (EQ) 1985 and 1987	Added additional reactor volume to the phenol and BPA production units.	Heavy ends (H, N)		
Solid (EQ)	Installed new heavy ends tower with distillation column at Phenol I unit to increase product yield.	Heavy ends (H)	17%	9,450,000 lb/yr
Solid (PS) 1985	Process changes resulted in a purer grade of phenol, reducing organic waste generation from BPA production.	Heavy ends (H, N)		
Water (OP) 1987	Reused hazardous wastewater for making raw material solutions at Phenol I unit.	Wastewater (H)		1,717,000 gal/yr (14,279,000 lb/yr)
Water (CH) 1989	Substituted a nonmetallic material for chromium used for corrosion resistance in cooling water.	Chromium (H)	100%	
Water (OP) 1984	Drain liquid from spent BPA process filters back into process instead of disposing as wastewater.	BPA (N)		85,000 lb/yr
Solid (PS) 1987	Process change in the wash step of the Phenol I unit.	Boiler slag deposits (H)	25%	
Water (CH) 1988	Substituted Safety-Kleen solvent for Dowclene	Dowclene (H)	100%	
Air/solid (EQ) 1988	Installed sample loops on product sampling purge lines to return the liquid to the BPA and phenol processes. A valve can be opened to obtain a sample when needed without exposing streams to the air.	BPA, phenol (H, N)		

Key to source reduction types: CH, chemical substitution; EQ, equipment change; OP, operational change; PR, product change; PS, process change.
A blank indicates that the plant did not provide information.

Change in Yield	Dollars Saved	Dollars Spent	Motivation	Comments	Time Needed for Implementation
	$50,000/yr		Result of employee suggestion.	Savings from increased yield.	
	$2,400,000/yr	$7,071,000; expected payoff is 3 yr	From a program to increase yields through increased reactor capacity.	Increased residence time in reactor increases purity of product, reducing waste.	
	$1,037,000/yr	$2,250,000		Engineering study showed that heavy ends could be heated and cracked to produce more phenol.	
			Process changes identified through the Aristech Total Performance (ATP) program.		
		$12,500	Result of employee suggestion.		
			NPDES wastewater permit limits for chromium were tightened.	No cost savings. Increased water treatment costs by 40-50%.	
+85,000 lb/yr	$32,047/yr	$3,697	Result of employee suggestion.	Practice instituted after ensuring that all contact with air could be avoided so that the fluid would not solidify.	
			Process change identified through the ATP program.	Process change improves yield, reducing boiler slag deposits when heavy ends are burned.	
		$30,000 to ventilate work areas	Dowclene was classified as hazardous by the Ohio EPA due to 1,1,1-trichloroethane.	Safety-Kleen is flammable so work areas had to be modified. It is recycled by vendor.	
		$120,000	To comply with new volatile organic compound regulations of the Ohio EPA.	Purge lines used to be opened to the air and the samples disposed of off-site.	

25 percent overall between 1984 and 1987, even though production increased about 15 percent.

Four of the source reduction activities reduced heavy ends, the organic waste made up of relatively large heavy molecules remaining at the bottom of distillation equipment after the product has been distilled from the reaction mixture. The plant was originally designed to burn virtually all heavy ends for energy recovery. The reduction in this waste over the last 5 years can be seen by the fact that natural gas use has increased 72 percent since 1984 and now supplies 80 percent of the fuel needs at the plant.

One of the heavy ends reduction activities occurred when an employee suggested, in 1984, that some of the heavy ends could be reduced by recycling the BPA mother liquor before the last step in the process, where it was burned off. The improved yield produced $50,000 in increased sales per year.

Another reduction in heavy ends was accomplished through a program to increase reactor capacity that was aimed at improving yields. Increased capacity in the reactors makes longer residence time by the chemicals possible, producing a purer grade of product. It also reduces the formation of organic wastes (heavy ends). First, in 1985, an additional "H" oxidation reactor was installed for the phenol operations at a cost of $786,000. In 1987, two additional reactors were installed in the BPA production unit at a cost of about $2 million. Four more reactors, costing a total of $4,285,000, were installed in the phenol production unit to improve reaction conditions and reduce waste generation. Despite the more than $7 million cost of these changes in reactor volumes, they paid off within about 3 years.

To process heavy ends for further product recovery, the plant added a heavy ends tower with a distillation column at the Phenol I unit at a cost of $2,250,000. An engineering study found that the large molecules of the heavy ends could be heated and broken down to produce more phenol. The distillation column has reduced the generation of heavy ends by 17 percent, from 90 to 75 pounds per thousand pounds of phenol, for a savings of $1,037,000 per year.

Plant officials reported further reduction in generation of heavy ends through process changes, made in 1985 as part of the ATP program, that resulted in production of a purer grade of phenol. When used as a feedstock for producing BPA, the purer phenol in turn reduces organic waste generation from that process.

Two of the reported source reduction activities affected wastewater. In one, in 1987, an employee suggested the reuse of wastewater for making raw material solutions at the Phenol I unit. The project cost a total of $12,500 and reduces hazardous wastewater by 1,717,000 gallons per year.

In addition, as a result of stricter NPDES wastewater permit limits imposed in 1989, Aristech replaced the chromium used for corrosion resistance in its cooling water with a newly available nonmetallic material. This has eliminated any chromium in the wastes from the cleaning cooling water tower (blowdown waste). There are no cost savings; in fact, waste treatment costs have increased by 40 to 50 percent.

Aristech reported four source reduction projects involving waste requiring off-site disposal. In one, an employee suggestion led to the draining of spent BPA process filters and putting the liquid back into the process instead of disposing of it as waste. This practice was instituted only after Aristech could ensure that all contact with air could be avoided so that the liquid would not solidify. This project cost $3,697 in 1984 and produces annual savings of 85,000 pounds of product, worth $32,047 a year.

In another, a 1987 process change in the wash step at the Phenol I unit (identified by the ATP program) has both improved yield and reduced boiler slag deposits when heavy ends from the phenol production are burned. Slag deposits have been reduced by 25 percent.

In 1988, Aristech substituted Safety-Kleen solvent for Dowclene in all its processes. Dowclene was classified as hazardous by the Ohio EPA because it contains 1,1,1-trichloroethane. Safety-Kleen is not potentially toxic. However, it is flammable, and areas where it is now used have been modified to provide good ventilation, at a cost of about $30,000. The new solvent is recycled by the vendor.

Aristech made a further change reducing solid waste from its BPA and phenol processes in 1988, in order to comply with new volatile organic compound regulations established by the

Ohio EPA's air office. Previously, when the company produced "purged samples" of its chemical products, (samples from lines purged of any old material so that they are fresh and representative of the current batch), the purge lines were opened to the air and the samples were disposed of off-site. Now, suction pumps and sample loops recycle the product directly back to the process through a closed system. A valve can be opened to obtain the needed sample at any time. Samples are put back into the process whenever possible. The cost to install the closed sample purge equipment on the BPA production unit was $120,000.

Overall, as indicated earlier, Aristech reported a 25 percent reduction in Haverhill plant waste while its production rose 15 percent between 1984 and 1988. Table II-3 shows the change in waste generation for each of these years, both overall and for wastewater, heavy ends, and off-site waste, reported as pounds of waste per pound of product.

Table II-3 Aristech (Haverhill, OH): Changes in Waste Generation, 1984-1987

Year	Production Index*	Total Waste (lb/lb of product)	Total Wastewater (lb/lb of product)	Heavy Ends Burned (lb/lb of product)	Off-Site Waste (lb/lb of product)
1984	1.00	0.60	0.55	0.045	0.00045
1985	0.85	0.65	0.61	0.045	0.00070
1986	1.05	0.55	0.50	0.035	0.00045
1987	1.15	0.45	0.41	0.041	0.00041

Ratio of the amount of product in one year compared to the amount of product in 1984.

Overall, waste as a fraction of the amount of product has decreased — from 0.6 pounds to 0.45 pounds (or 25 percent). The majority of the waste is wastewater, which has also decreased 25 percent, from 0.55 pounds per pound of product to 0.41 pounds per product. Heavy ends and off-site waste each decreased by 9 percent as a fraction of the amount of product.

While the trend has clearly been toward waste reduction, increases in waste generation occurred in 1985. This was due to heavier rains than normal, increasing the total wastewater, and to the removal of BPA sludge from the plant's surface impoundment, which used to be done every 3 to 4 years. This pond, however, was closed in November 1988. BPA waste is now treated above ground with a clarifier and filter press.

Other Waste Management Practices

Aristech reported three waste management practices other than source reduction activities to INFORM.

Plant officials, in 1987, found ways to reduce the number of expensive filter cartridges used as polish filters for the plant's injection well. They substituted inexpensive bag filters, where possible; the smaller size of these filters reduced the number of drums needing to be disposed of as hazardous waste. The cartridges last three times as long with the new filters, also reducing the number of drums needed. The cost of the change was $17,800. The use of filter bags reduces the volume of the waste collected without affecting its overall weight and prolongs the life of the more expensive cartridge filters.

In the area of wastewater, an employee suggestion led to the separation of clean wash water from the hazardous process wastewater for AMS washes at the Phenol I unit. There was no cost for implementing this change, which led to a reduction of 965,000 gallons per year going to the injection well. Another employee suggestion led to a second instance of segregating nonhazardous wastewater, this time cooling water flows, at a cost of $859 and reduction of 500,000 gallons per year of wastewater that had to be managed as "hazardous waste," saving $3,330 per year. A third project segregates nonhazardous compressor drains from hazardous wastewater drains at the Phenol I unit, costing $9,850 for a reduction of 1,275,000 gallons per year in wastewater that must be managed as "hazardous waste." These suggestions were implemented in 1986.

When the RCRA amendment defining stricter design controls on secondary containment structures, such as storage tanks, was promulgated, Aristech found that its oil separation tank needed a new liner. However, further study determined that a coalescer would do a better job of containment and allow oils to be reclaimed and reused. A new coalescer is being installed at the phenol waste treatment plant at Haverhill. The coalescer will reduce the toxicity of the phenol wastewater and allow reclaiming and reuse of the oils in the wastewater.

Technical Assistance

According to company officials, the formation of Aristech and its adoption of its ATP program has greatly increased cross-communications among plants in this company. Information is shared more in meetings and telephones calls than in formal written communications. For example, representatives from all plants with boilers meet to discuss boiler problems. In addition, personnel from other plants are members of review teams at the Haverhill plant, and vice versa.

Plant personnel get technical information from trade publications and attend seminars sponsored by the Ohio Manufacturers Association. They evaluated the OMA seminars as very helpful, especially when the sessions are attended by government officials who can interpret regulations for them.

The Haverhill plant has listed wastes that could be available to recyclers on mailing lists for recyclers. So far, however, the plant has received no applications for its wastes.

Company Comments on State and Federal Regulations

The company indicated that delays in reviewing and approving plants' applications have hindered introduction of some source reduction projects. One example is the Haverhill plant's application to change from the use of chromium to a nonmetallic material in its cooling water. Aristech's application was on file for over 9 months before it was approved and the plant could implement the change. Plant officials reported a similar problem with its RCRA applications. Part A of the Haverhill plant's RCRA permit was approved in 1981; several amendments have been filed since then, including one to change the name to Aristech. However, mail and government inspections are still based on the USS Chemicals 1981 application. A further complication, according to the company, is the fact that any actions by the Ohio EPA must be approved by the federal EPA because Ohio has not been delegated the administration of some programs. Aristech reported that Ohio EPA has also been lacking in resources to run its programs, causing further delays.

Aristech also reported that from 1980 to 1982, while RCRA amendments were being debated, no one knew what changes would be required, so they could not plan. Consequently, they faced possible delays in response to the changes since engineering studies can take 5 years or more. Plant officials believe that if, instead, there had been a cooperative, consensus process allowing industry to participate in the outcome, they could have planned ahead and begun the studies earlier.

Haverhill plant officials view the requirements of Section 313 of the Superfund Amendments and Reauthorization Act (SARA 313) in two ways. First, they stated that something like this was needed to cause the plant to draw together waste and emissions information that was available, but not readily accessible in a single concise document. However, the plant manager is anxious about community interpretation of these numbers, believing they are not presented in context. The Haverhill plant has been participating in the Community Awareness and Emergency Response (CAER) program of the Chemical Manufacturers Association, which stresses local emergency response planning and public education about chemicals, and plans to increase its efforts as the SARA 313 information becomes available. On Aristech's initiative, Haverhill plant personnel have met with officials of the three counties surrounding the plant to assist them in emergency planning.

While not a problem for the Haverhill plant, officials also said the Toxics Release Inventory (TRI, part of the SARA 313 requirements) database has the potential to hinder research into proprietary processes that might result in source reduction. They believe that the data required by TRI could give other companies sufficient information to duplicate processes or portions of them, and that companies may not want to spend the large resources necessary for such process development if they cannot have exclusive rights to it so that they can recover the research costs.

Future

The largest waste management issue facing the Haverhill plant involves its use of injection wells. The RCRA 1984 amendments called for a ban on such wells unless a petition for exemption was approved by August 1988 (there has since been a blanket extension). There has been no action by EPA on what the regulations for this will be. Once the regulations are promulgated, the plant manager pointed out that the plant will have to undertake studies of process improvements to reduce these wastes and that such studies always take time.

ATLANTIC INDUSTRIES
Nutley, New Jersey

Summary

Atlantic Industries is a large dyestuff manufacturer located in the town of Nutley, New Jersey (a residential area of northern New Jersey). The plant, built in 1939, has 240 employees and produces hundreds of different dyes in batch operations. Raw materials include aniline, phenol, chromium, and formaldehyde. It is managed by a small team of four people who, in addition, share responsibility for all environmental aspects of the plant operations. They do not have written policies and have concluded that batch operations such as theirs are not amenable to source reduction by means of tracking materials in the waste back to their source. Source reduction has primarily resulted from yield improvements through process changes. Plant officials also mentioned an informal goal of discontinuing the use of chemicals that raise health and safety concerns, either ceasing production entirely or substituting a less hazardous chemical, if possible. The primary types of waste at this plant are air emissions, which are treated with scrubbers, and wastewater discharges to the local sewage treatment plant, also treated when necessary.

Atlantic granted an on-site interview to INFORM and conducted a tour of the Nutley, New Jersey facility. The plant reported a total of seven source reduction activities since the publication of INFORM's *Cutting Chemical Wastes*. Atlantic had also granted an interview for that 1985 study, and at that time had reported two other source reduction activities, reducing 350,000 pounds of waste each year.

Products and Operations

The Atlantic Industries Nutley plant is a large dyestuff manufacturer, with 240 employees. Some 200 to 300 different chemicals are produced every year at the plant, exclusively in batch operations. Originally, this plant registered about 750 to 1,000 chemicals under the Toxic Substances Control Act (TSCA), but relatively few of these are produced over the course of a year, and most are produced in amounts of less than 25,000 pounds per year.

From 1984 to 1987, Atlantic's volume of production increased by 15 to 20 percent, with sales growing to $60 to 70 million a year. During 1985, the company closed two Atlantic warehouse sales offices and opened a new one, in Greenville, North Carolina. The Greenville plant is not a manufacturing facility; it is used solely for blending operations and as a warehouse, customer service laboratory, and sales office, which has freed up some space at the Nutley plant for increased production.

Atlantic Industries underwent two changes in ownership during the second half of the 1980s. In 1985, Jepson Company, Inc., a holding company, bought the Atlantic Chemical Corporation of which Atlantic Industries is a subsidiary. Jepson is an investment concern, and the other companies it owns are not dyestuff manufacturers. This change in ownership subjected the Nutley plant to provisions of New Jersey's Environmental Conservation and Recovery Act (ECRA) which require that a condition for sale of property is that it be verified to be uncontaminated. An analysis of soil and groundwater at the plant site identified some problems and a remediation plan for these was approved; implementation began shortly thereafter.

In 1989, Jepson sold the plant to Great American Management, an investment company which put Atlantic up for sale again in 1990. These changes in ownership did not have any affect on plant operations: there were no changes in personnel and the plant managers still make operational decisions.

Environmental Policy

Atlantic Industries prides itself on its informal management style. Not only does it not have a written environmental policy, but until recent years it also had no written budget or organization chart. A group of top managers monitors the plant operations and relies on the expertise of the individuals of the group to point out problems. The primary responsibility for all activities related to environmental regulations at the plant is shared by the company's vice-president—technical director, senior vice-president of engineering, vice-president—general manager, and plant engineer.

These managers reported that this approach is working well for their plant, noting that, over the past 20 years, the president and top managers have established a stewardship toward the plant employees and the surrounding residential community that has achieved environmentally sound practices (as witnessed by the lack of major problems found under the ECRA review), as well as high unit sales and production for this size and type of plant.

Materials Data Collection

Atlantic Industries does not have a materials or cost accounting system. With hundreds of chemicals produced in relatively small quantities, and with some chemicals produced just once in 2 or 3 years, the company does not track wastes back to their source because of the constant change in sources. According to the plant's research, reducing waste in a given operation is generally not cost-effective because each batch operation is short and is a relatively small part of total production.

Other Source Reduction Program Features

Of the remaining key source reduction program features outlined by INFORM, Atlantic Industries has partially adopted source reduction leadership, but does not set numerical goals for source reduction or have employee involvement programs.

While no formal source reduction program exists, Atlantic reports a continuing program of process improvement with one of its aims being reduction of waste. This program aims to improve the efficiency of the processes by which the chemicals are produced, with resulting reductions in waste as a by-product of these endeavors.

In response to US EPA regulations requiring that the organic chemical industry pretreat waste before sending it to sewage treatment plants, Atlantic has undertaken a program to look at biological oxygen demand (BOD) and priority pollutants (126 specific federally listed toxic chemicals in wastewater discharges) in their wastewater stream. Plant officials expect that, in order to meet these standards, the plant will have to treat the wastes rather than being able to reduce them at source because of the highly variable nature of their operation.

Source Reduction Activities

Atlantic Industries reported implementing seven source reduction activities to INFORM for this study. The activities are summarized in Table II-4 and described below. Two earlier source reduction activities are described in INFORM's 1985 report.

Three of Atlantic's source reduction activities were motivated by a desire to improve yields. In one, changes made in crystallization time (a purification process) resulted in a 50 percent reduction in the amount of wash water used for the particular process involved. This meant that less product loss also occurred, since both product and impurities are inevitably lost in rinses. Decreased washing also means less wastewater and lower BOD levels in the plant effluent discharged to the Passaic River Valley sewage treatment plant.

A second example of source reduction as a result of process improvement made mainly for increased product yield is Atlantic's development of processes that make it possible to mix all chemical components for a dyestuff in one kettle. Dye-making previously required that materials be filtered, rinsed, and re-worked; when filters were used, some of the product ended up in the rinse water and then in the sewer. With the final product made in one kettle without filters and rinses, cost savings can be as much as 10 percent or occasionally more of the manufacturing costs. However, the impact of this process change on overall production has been small, so far, because it has been applied to only 1 to 2 percent of the production processes.

A third process improvement aimed at increasing yields involved changing the concentra-

Table II-4   Atlantic Industries (Nutley, NJ): Source Reduction Activities

Waste Medium (SR Type) Year	Source Reduction Activity	Specific Waste Reduced (Hazardous or Nonhazardous)	Percent Waste Reduced	Amount Waste Reduced
Water (PS) 1984-1987	Changed crystallization time for some products.	Reaction vessel residues BOD (N)		
Water (PS) 1984-1987	Developed liquid processes that are conducted in-situ. Dyes are produced in a single kettle with no need to filter or rinse.			
Water (PS) 1984-1987	Changed relative concentrations of reagents in order to drive reactions further to completion and increase yields.			
Water (PR) 1984-1987	Search for replacement products. Expect 2 or 3 approved under the premanufacture notification process of TSCA.			
Water (OP) 1984-1987	Eliminated storage of oleum in tanks due to risk of spills. Oleum is now only used in drums.			
Water (PS) 1984-1987	Searching for ways to increase concentration of final product so it can be spray dried instead of precipitated, isolated, and filtered.			
Solid (OP) 1984-1987	Changed from fiber drums to plastic drums, which can be reused, for internal use.	Trash (N)	33%	

Key to source reduction types: CH, chemical substitution; EQ, equipment change; OP, operational change; PR, product change; PS, process change.
A blank indicates that the plant did not provide information.

tion of reagents in order to drive the chemical reactions further to completion.

Two source reduction activities were motivated by Atlantic Industries' continuing informal program goal of eliminating use of chemicals that could cause health or environmental problems because of company concern for its residential neighbors and employees, as reported in INFORM's previous report. In one activity, Atlantic has been looking for replacement products. These must be approved by EPA's Office of Toxic Substances; when the plant submits a required premanufacture notification (PMN) to EPA, EPA has 90 days to object or production can begin. These new products would likely represent less than 1 percent of production, and Atlantic expects that production costs will increase because of the use of the substitute chemicals.

As an accident prevention measure, reflecting Atlantic's concern that it live responsibly and at peace with its residential neighbors, Atlantic has eliminated the use of certain kinds of chemicals

Change in Yield	Dollars Saved	Dollars Spent	Motivation	Comments	Time Needed for Implementation
			Increase yields and improve productivity.	Fewer rinses reduces washings and lost product; 50% reduction in water use.	
	10% or more of manufacturing cost		Increase yields and improve productivity.	Overall impact small; just 1 or 2% of production processes have been affected.	
			Increase yields and improve productivity.		
			Program to eliminate use of chemicals that could cause health or environmental problems.	Change likely affect less than 1% of production. Costs could increase.	
			Concern that Atlantic live responsibly and at peace with the plant's residential neighborhood.		
			Reduce energy costs.	Wastewater as well as energy costs would be reduced.	
			Quadrupling of cost of nonazardous trash removal.	Overall costs have gone up due to increased removal costs.	

and handling and storage operations. For instance, the plant no longer handles bulk oleum (concentrated sulfuric acid). Oleum is only used in drums because of the risk of spills from tanks.

A desire to reduce costs motivated two source reduction activities. In one, aimed at reducing energy costs, Atlantic Industries is looking into ways to increase the concentration (i.e., decrease the amount of water used) in the final product solution in the kettle so that products can be sprayed dry instead of having to be precipitated, isolated, and filtered. This process change would reduce wastewater.

The cost of nonhazardous trash removal has quadrupled in New Jersey in recent years. At Atlantic, such trash consists of fiber drums, paper bags, and computer paper, and the costs of its removal amounted to approximately $250,000 in 1987. Atlantic pays to have the trash removed because there is not enough to collect and sell; a full trailer would have to be collected first and

Atlantic's operations are too small to make that feasible. Atlantic investigated building an incinerator but decided against trying for this option since only four or five had been licensed in New Jersey in the preceding 5 years. Hence, Atlantic turned to source reduction and has reduced its volume of trash by about one-third by replacing fiber drums with plastic reusable ones. However, its costs have gone up because of the overall increase in trash removal costs.

Other Waste Management Practices

In addition to its source reduction activities, Atlantic reported three waste management practices. One problem Atlantic has experienced is that wastewater is occasionally more acidic than allowed by the sewage treatment plant to which Atlantic discharges this wastewater. Larger plants build ponds to mix effluent streams to neutralize the acidic loads, but Atlantic's site does not allow for this. The company first tried to monitor the different wastestreams so that they could be mixed. However, this approach did not result in the wastewater's acidity falling within the parameters (pH 5.5 to 10.5) required by the sewage treatment plant. Instead, Atlantic has been able to control pH by approaching the problem as a process problem, rather than a waste-handling problem. Now every possible process is controlled so that the final product wastestream is adjusted to near neutral pH. For those few effluent streams for which this is not possible, the acidic solution is stockpiled until it can be mixed with a basic stream. It should be noted, however, that in adjusting the pH load to conform to the strict regulatory requirements (only a 1 percent deviation, the equivalent of 15 minutes in 24 hours, is allowed), the inorganic load in the plant's effluent (in the form of sodium sulfate) has increased.

Almost all of Atlantic's waste is wastewater, which is monitored for BOD and suspended solids as required by the Passaic River Valley sewage treatment plant. As new regulations on pretreatment are passed, the plant will monitor for such chemicals as chromium and copper sulfates, reusing them in batches where possible or otherwise precipitating them out of the wastewater. All storm water run-off is collected into the wastewater stream, which is discharged to the sewage treatment plant. All air emissions are scrubbed, and solid wastes are mostly paper and empty containers.

A lot of Atlantic's raw materials from overseas come in steel containers. There used to be a thriving business in recycling these drums (the drums were washed, painted, and resold), but Atlantic can no longer find recyclers because of the fear of liabilities in their reuse. Thus, Atlantic's disposal costs have risen because it must triple-rinse the drums, crush them, and certify that they are not a source of contamination before disposal. The problem remains that soon the drums may be rejected entirely by landfill operators.

Technical Assistance

The plant managers reported that, by necessity, they have to be knowledgeable about environmental regulations. They said that it would be a great help if government provided industry with assistance in eliminating the use of specific chemicals when it promulgates a rule that these chemical wastes must be eliminated.

Most of the technical information Atlantic has used to replace products or reagents has come from general literature on current technology.

Company Comments on State and Federal Regulations

Plant officials state that the largest impact government regulations have had on the plant has been to require so much paperwork that time cannot be spent on looking for ways to reduce wastes. Atlantic managers assert that much of the information asked for is not useful to them. They also say that the EPA could do a better job in deciding which chemicals are placed on lists of hazardous chemicals so that companies do not spend time accounting for chemicals that are later removed from the lists.

Atlantic is participating in a group of chemical industry officials that is working with EPA on a major "waste minimization" program for the entire industry.

BONNEAU DYE CORPORATION
Avon, Ohio

Bonneau Dye Corporation is located in Avon, Ohio, about 15 miles west of Cleveland. It manufactures custom blend dye formulations and speciality chemicals for candle-making and other crafts. Bonneau did not cooperate in INFORM's 1985 study and did not grant an interview for this one. A total absence of data in public records precluded analysis on Bonneau's waste generation and handling practices. There is no indication from available information that Bonneau has adopted any source reduction techniques.

BORDEN CHEMICAL COMPANY
Fremont, California

Summary

The Borden Chemical plant in Fremont, California, an industrialized area southeast of San Francisco, is a medium-sized plant with 50 employees. Built in 1959, it manufactures adhesives and resins for use in other industrial processes. In 1987, the Borden Fremont plant ceased production of all contact cement.

As a result of strict regulation of phenolic and formaldehyde discharges in the wastewater going to the municipal sewage treatment plant, Borden has undertaken systemic, plant-wide studies of its waste and its sources, and has achieved over 95 percent reduction in total wastes through source reduction. Further, state and federal regulations of disposal options (landfills) have also led the plant to investigate further source reduction measures; these have turned out to have large economic benefits, with paybacks of a year or less for most projects. With this plant leading, the overall Borden corporation has recently developed a corporate environmental policy manual that includes a section on source reduction.

Borden granted an on-site interview to INFORM and conducted a tour of its Fremont facility. The plant reported seven source reduction activities, reducing 293,070 pounds of waste and saving the company $46,620 each year. Borden had also granted an interview for INFORM's 1985 study and at that time reported six other source reduction activities that saved the company $48,750 each year.

Products and Operations

The Borden Chemical Company is a part of a large international corporation, Borden, Inc., and represents 16 percent of the corporation's total revenues. The Fremont plant is a part of Borden's Adhesives and Resins Division. It manufactures formaldehyde, urea-formaldehyde, phenol-formaldehyde resins, and wax emulsion sold for use in industrial processes. The level of production and employment (about 50 employees) has remained about the same since 1983. The Borden Fremont plant ceased production of all contact cement adhesives in 1987.

Environmental Policy

Borden does not have a written policy favoring source reduction over other waste management options. However, in early 1989, a corporate-wide environmental manual was issued to all Borden facilities, providing details on environmental policy and practices, including comprehensive sections on source reduction and energy conservation. In 1987, Borden launched an official "waste minimization program," in which process engineers from the corporate environmental affairs department go to individual plants and work with plant personnel to search for ways to improve plant processes in order to reduce wastes.

The Borden corporation encourages individual plants to find ways to meet local and state regulations through source reduction and other waste management techniques because local and state laws differ and equipment and product lines can vary greatly among the plants. Environmental and safety engineers, new positions established in 1987, have been hired for Borden's Adhesives and Resins Division (of which the Fremont plant is part) and meetings between the division head and the plant managers have emphasized Borden's desire for zero discharge of wastewater. While source reduction at the plant is primarily overseen by the plant manager, this strategy is also considered one of the responsibilities of the environmental and safety engineers.

Other evidence that Borden corporate management is putting greater stress on environmental practices at its plants is the required environmental assessment for each plant. This environmental assessment entails an audit by an outside risk management consulting firm and an audit by corporate and plant environmental and safety engineers. The Borden Fremont plant completed its first assessment in 1987.

An incentive plan for hourly employees at the Fremont Borden plant, in place since 1985, includes source reduction. The plant manager sets goals for the number of drums of waste that can be generated in different parts of the plant. Employees can lose part of their bonus if the reduced waste generation goal is not met.

Materials Data Collection

Borden has established partial materials accounting systems at the Fremont plant. For the formaldehyde operations, yield is tracked daily to monitor the efficiency of the catalyst. A complete materials balance for the formaldehyde operations has not been done because of the analytical complexity of measuring all reaction by-products. However, the odor of formaldehyde is so pungent that even small amounts of leakage are noticeable, and formaldehyde leaks must be promptly repaired in order to comply with personal exposure limits established by the Occupational Safety and Health Administration. Borden has also installed total organic carbon analyzers on its wastewater system. However, it has not yet had success in having the instruments work reliably. Daily inventories and analysis are performed on Borden's wastewater storage tanks to track the volume and strength of the wastewater being released.

Other Source Reduction Program Features

Of the remaining key source reduction program features outlined by INFORM, Borden has fully or partially adopted cost accounting, leadership, employee involvement, and numerical goals programs.

The costs of wastes are allocated to each product line at the Borden plant. Costs of all types of wastes are included and are based on waste categories, rather than individual chemicals. Specific costs of wastes under this allocation include lost materials, disposal, regulatory compliance, insurance, and other liabilities, such as accidents and clean-ups.

Source reduction at the Fremont plant is primarily the result of the plant manager's leadership and his employees' knowledge of the processes. The source reduction program has been undertaken in response to ever stricter government regulations, particularly limits on chemicals in the wastewater discharged to the municipal sewage treatment plant, but also federal and state regulation of underground storage tanks, landfills, and other solid waste disposal. The plant manager reported that regulations provide the incentive to reduce waste and source reduction has proven to have economic as well as environmental benefits.

A step has been taken toward institutionalizing source reduction at this plant through the formation of "waste reduction teams" to focus on particular problems. The first problem for which a waste reduction team was formed was formaldehyde in the wastewater.

The Borden Fremont plant also has begun to apply statistical process control (SPC) techniques to its processes, which has greatly helped the waste reduction team. By doing their own measurement and analysis, operating officials have the control over processes at the point where control actions can be taken most efficiently. The data from SPC are process-specific and can isolate major sources of waste; they are based on individual chemicals and include nonprocess areas, such as loading/unloading and other materials transfers. Currently, SPC data measure wastes released to water only.

The general manager of Borden's corporate Adhesives and Resins Division has set a goal of zero discharge of wastewater. Currently, the focus is on formaldehyde and phenol in wastewater at the Fremont plant; however, air emissions and solid wastes are also targets for source reduction. Quarterly reports are being sent to the division's engineering and environmental staff on the plant's progress towards achieving zero discharge of phenolic wastewater.

Source Reduction Activities

INFORM's 1985 study describes the six source reduction activities that Borden reported at the time; for this new study, Borden reported seven more which are summarized in Table II-5 and described below.

Four of Borden's source reduction activities were aimed at reducing the discharge of formaldehyde in wastewater. This became a focus of attention in 1987 when the local sewage treatment plant imposed a new limit of 50 parts per million (ppm) on the wastewaters it would accept; previously, there was no limit. In all, the formaldehyde content in the total plant sewer wastewater being discharged has been reduced by 95-100 percent, over 280,000 pounds a year. In addition the volume of wastewater discharged has been reduced by 3,000 gallons a day. The projects cost a total of about $36,000 and have produced waste reduction savings of over $25,000 a year.

Most of the source reduction measures used to reduce formaldehyde in Borden's wastewater took about 6 months to implement. The plant manager reported that plant operators already knew what to look for because of the plant's experience in reducing phenol (as reported in INFORM's 1985 study), making the process "surprisingly easy." The application of statistical process control (SPC) techniques to Borden's operations also helped to get projects moving.

One set of formaldehyde-reducing steps utilized closed-loop recycling of formaldehyde-contaminated seal waters. Vacuum pump seal water in the urea resin unit is recycled as water for the formaldehyde process. Also, when the water in the seal water reservoir for the phenolic resin reactors is replaced, the spent water is recycled back into the resin manufacturing process. This activity also involved replacing carbon steel vacuum pumps with stainless steel ones that do not corrode as the concentrations of formaldehyde rise. The stainless steel pumps cost $6,000 more than the carbon steel ones. The replacement began in 1988 and eventually there will be three pumps in use. Total costs for this project were $20,000 and the expected savings are $420 per year. The formaldehyde wastes have been reduced by 6,000 pounds per year or by 98 percent. Increases in yield were small, amounting to 0.005 percent.

In 1987, to reduce another source of formaldehyde contamination of wastewater, filter wash water from resin manufacturing began to be reused in the process. The main filter used in truck loading of urea-formaldehyde resin is commonly back-flushed three times a day. The rinse water from the first two flushes is now caught and returned to the product. The excess water is distilled off in the production of urea-formaldehyde resin. The distillate is reused as make-up water in the formaldehyde process and residual dissolved resin is recovered. A total of 150,000 pounds per year of formaldehyde wastes have been eliminated, for a 95 percent reduction for this wastestream. In addition, this change increased yield by 0.2 percent and contributed 30 percent in the reduction of the plant's chemical oxygen demand (COD). The costs for the equipment to catch and return the rinse water to the process were $2,000, while annual savings have been $14,250 in avoided sewer and sludge disposal fees.

The entire area where formaldehyde is manufactured has also been isolated by trenches and dikes. In this way, leaks and equipment rinses can be collected. This wastestream is reused in the process if the analysis at the collection tank shows the level of impurities to be acceptable. One hundred percent of the formaldehyde wastes from this source have been eliminated, and the change has contributed to a 10 percent reduction in COD in wastewater for the plant as a whole. This action cost about $10,000 for the pipes and tank. There has been a small improvement in yield (0.05 percent), but the significant cost savings have been in compliance with the formaldehyde discharge limit and reduction of COD in the wastewater going to the sewage treatment plant. As a result, sewage treatment plant fees have been reduced by $2,000 per year, and overall savings from this project have been $3,500 per year.

Yet another formaldehyde reduction change was made in 1987 — a change in the way process filters used in the manufacture of urea-formaldehyde resin are cleaned. Before 1987, the particles were flushed in wastewater to the sewer. Now, the particles are collected in drums for disposal as hazardous waste, and the rinse water is collected in Tote bins and reused in the resin manufacturing process. The cost of the drums and Tote bins was $4,000 and the savings are $7,200 per year. This change cut the formaldehyde waste from this operation by 95 percent (75,000 pounds a year), increased yield by 0.1 percent, and contributed 15 percent of the plant's total COD reduction.

The rising costs of hazardous waste disposal motivated two source reduction activities. In one, Borden plans to close the pits it had been using for collecting wastewater rather than upgrading them to meet California surface impoundment specifications. These concrete pits, used to settle out sludge from wastewaters, with the sludge then being sent to a hazardous waste landfill for disposal, could have been classified as "hazardous" waste disposal sites as defined under the Resource Conservation and Recovery Act and would have been subject to expensive regulation.

As a result of the decision to close the pits, Borden has completely redone the Fremont plant's wastewater system. Since 1987, Borden has spent $250,000 on a new above-ground wastewater collection system and central sludge holding tank. Plant officials expected that this holding tank would have to be vacuumed four times per year to remove the sludge. However, the source reduction measures taken in the urea-formaldehyde area also reduced sludge generation by 80 percent, for savings of over $17,750 per year in avoided sludge disposal costs.

The disposal cost of hazardous wastes has also given Borden an incentive to find ways to reduce wastes that would otherwise require disposal. The Fremont plant is now using reusable Tote bins rather than nonreusable drums to ship its products to its customers, and also requires its suppliers to use them. The plant manager finds that it is difficult to control their use and, thus, to clean them without generating a lot of wastewater, but they are popular because of the problems of disposing of nonreusable drums. This change has reduced the need to dispose of empty drums by 90 percent, or 300 drums per year, and saves $2,000 per year.

In its seventh reported source reduction activity, Borden achieved further reductions in phenol waste beyond the series of operational changes described in INFORM's 1985 report that reduced phenol waste by more than 90 percent. These further reductions have been accomplished by collecting the isolated wastewater in the trench coming from the tank house and the phenol rail car unloading area in a 30,000 gallon surplus tank. This "clean" water, containing only phenol and formaldehyde, is put back into the resin batches. Before 1988, the water went to an ozone treatment unit. Currently, the ozone unit is needed only during the rainy season when storm water run-off is high. Waste, formerly 70 pounds a year, has been completely eliminated. Although a somewhat smaller yield is the result, $1,500 per year is saved in treatment costs. The project cost $3,000.

Other Waste Management Practices

In addition to these source reduction activities, the Borden corporation consolidated the production of contact cement, resulting in discontinued production of solvent cements at this plant, in order to eliminate the need for underground tanks to store solvents used in producing contact cement adhesives. California law permits such tanks, with annual test for leaks, but the potential for problems with such tanks led to the company's decision to eliminate their use. As a result, Borden is eliminating the need to dispose of approximately 30 drums of RCRA wastes per year from the Fremont plant. The drums contained samples, wash solvents, and off-quality products. This was the only source of RCRA wastes at this plant, although some of its waste sent off-site is regulated under California law.

Technical Assistance

The Fremont plant manager said he has not used the Alameda County Health Department phone number for information on source reduction practices and so does not know the quality of the information available.

In 1985, the Chemical Manufacturers Association initiated a Community Awareness and Emergency Response (CAER) program to encourage chemical manufacturing facilities to improve safety and performance and to INFORM local residents about industry operations and emergency response plans. Some CAER groups have existed for several years, but many are being set up in response to Title III of the Superfund Amendment and Reauthorization Act (SARA). This law requires industry to INFORM communities about the types of chemicals it uses and releases and about its emergency response plans. While there is no CAER group in Fremont, neighboring Newark does have one. The plant engineer has attended some of the meetings and

Table II-5   Borden Chemical Company (Fremont, CA): Source Reduction Activities

Waste Medium (SR Type) Year	Source Reduction Activity	Specific Waste Reduced (Hazardous or Nonhazardous)	Percent Waste Reduced	Amount Waste Reduced
Water (OP, EQ) 1988	Recirculate formaldehyde-contaminated vacuum pump seal water back to process. Carbon steel pumps replaced with stainless steel pumps. Similarly, contaminated reaction seal water is recirculated in phenolic resin process.	Formaldehyde (H)	98%	6,000 lb/yr
Water (OP) 1987	Reuse flushings from truck loading filters. Filtrate from the first two flushes is returned to the process. The excess water is distilled off and reused in process.	Formaldehyde (H)	95% (30% of chemical oxygen demand [COD] reduction for total plant)	150,000 lb/yr
Water (OP) 1987	Isolated the formaldehyde manufacturing unit with trenches and dikes to collect wastewater from leaks and equipment rinsing. Collected water is reused in process.	Formaldehyde (H)	100% (10% of COD reduction for total plant)	50,000 lb/yr
Water (OP) 1987	Modified product filter cleaning operation in urea-formaldehyde resin manufacturing process. Water is collected in Tote bins and reused in process	Formaldehyde (H)	95% (15% of COD reduction for total plant)	75,000 lb/yr
Solid (OP,EQ)	Reduced wastewater treatment sludge generation through formaldehyde-related activities.	Sludge (H)	80% reduction for total plant	
Solid (EQ) 1987	Package products in reusable Tote bins rather than nonreusable drums. Borden requires its suppliers to do the same.	Empty drums (H)	90%	300 drums/yr (12,000 lb/yr)
Water (OP) 1988	Collect and isolate wastewater in the trench coming from the tank house and phenol railcar unloading area in a 30,000-gallon tank. The water is reused in the resin batches.	Phenol (H)	100%	70 lb/yr

Key to source reduction types: CH, chemical substitution; EQ, equipment change; OP, operational change; PR, product change; PS, process change.
A blank indicates that the plant did not provide information.

Change in Yield	Dollars Saved	Dollars Spent	Motivation	Comments	Time Needed for Implementation
+0.005%	$420/yr	$20,000	Local sewage treatment plant imposed a formaldehyde limit of 50 ppm.	Stainless steel pumps cost about $6,000 more than the carbon steel ones and do not corrode as concentration of formaldehyde increases. Three new pumps needed.	0
+0.2%	$14,250/yr	$2,000	Local sewage treatment plant imposed a formaldehyde limit of 50 ppm.	The main filter used in truck loading is commonly back-flushed three times a day.	1 person-month
+0.05%	$3,500/yr	$10,000 for pipes and tanks	Local sewage treatment plant imposed a formaldehyde limit of 50 ppm.		0
+0.1%	$7,200/yr	$4,000	Decision to discontinue underground storage tanks.	Particles are collected in drums for disposal.	1 person-month
	Additional $17,750/yr	$250,000	Rising costs of hazardous waste disposal.	Savings are in sludge removal costs and reduced sewer charges.	6 mo
	$2,000/yr		Problems with drum disposal.	It is difficult to control use of the Tote bins and thus to clean them without generating wastewater, but they are popular because of the problem of disposing of drums.	
Negative	$1,500/yr in treatment costs	$3,000	Corporate goal of zero discharge of phenolic wastewater.	Before, the wastewater went to an ozone treatment unit. Now this unit is only needed during the rainy season.	

said he found it useful to meet with other people in the chemical industry to discuss problems. However, he added that it would be more helpful if the city of Fremont had its own group.

Each Borden plant receives environmental and safety incident reports from other Borden plants; these can contain information helpful to other plants' operations. Source reduction practices are also discussed at Adhesives and Resins Division meetings of plant representatives. The Fremont plant manager reported that the Borden Montana plant, for example, has reduced its wastewater discharge to zero, in part by adopting measures taken at the Borden Fremont plant. The Fremont plant, in turn, adopted some of its new waste reduction concepts from the Montana plant design.

Future

At the Borden Fremont plant, the next major area of focus for source reduction, after formaldehyde, will be in the resin operations. A waste reduction team has already been formed to achieve the goal of zero discharge of phenolic wastewater.

The area of air emissions may get more attention as it is becoming a new focus for the local government regulators. They have asked the plant manager to count the number of valves and flanges at his plant in order to estimate fugitive emissions for phenol and formaldehyde; both of these materials have been declared toxic air contaminants by the California Air Resources Board. The plant manager stated that this will be useful to him in filling out the plant's Toxics Release Inventory report. However, he sees problems if this count is used for estimating air emissions for other state and federal report forms. The usual practice is to apply a factor for each valve and flange to calculate emissions. However, the factor EPA uses is based on oil refinery operations, where lightweight hydrocarbons are predominantly used. Lightweight hydrocarbons of an oil refinery are different from the 50 percent solution of formaldehyde produced at this plant. Formaldehyde solution turns into a white solid polymer when exposed to air, making any formaldehyde leaks readily identifiable. At the Borden Fremont plant, the plant manager reports that emissions of formaldehyde are not high enough to be smelled, nor are white solid polymers forming around the flanges or valves, indicating leaks. Thus, he believes that using the EPA factors will result in very high emissions estimates that most likely would not accurately reflect actual emissions at the plant.

CHEVRON CHEMICAL COMPANY
Agricultural Chemicals
Richmond, California

Summary

The Chevron Chemical plant in Richmond, California, 10 miles northeast of San Francisco, is a large facility with 274 employees manufacturing agricultural chemicals, including pesticides, herbicides, and gasoline additives. This plant is sited near, but is wholly separate from, a large complex that includes a Chevron USA oil refinery, a research center, and a Chevron Chemical Fertilizer Division plant.

Spurred by overseas competition, the costs of raw materials, and growing public concern, Chevron established a written environmental policy favoring source reduction in the mid-1980s. The program established to implement this policy includes materials tracking procedures, cost accounting, and recognition of ideas from employees.

The California Department of Health Services, Toxic Substances Control Division, told INFORM that "all Chevron Chemical Company reports have been designated as trade secrets which makes them unavailable to the public." Chevron told INFORM that all company reports submitted to the government, except Toxics Release Inventory Form Rs, are designated as trade secrets because of concerns over competitive advantage due to the nature of its products.

In 1983, Citizens For a Better Environment (CBE) — California initiated a campaign targeted at the nearby Chevron oil refinery. A study, "Toxics in the Bay," conducted by the organization, revealed the refinery as the single largest discharger of toxic wastewater pollutants into the San Francisco Bay. For the next few years, CBE and Chevron were entangled in legal battles over the facility's exemption from Clean Water Act requirements. In 1986, the two groups entered into a dialogue that both called "constructive;" it culminated 9 months later in a cooperative agreement, announced at a joint press conference, that resulted in reduced Bay discharges.

The campaign over wastewater issues at the refinery added to longstanding health and safety concerns of the adjacent community. Currently, CBE is working with a local community group, West County Toxics Coalition, to address issues at the chemical facility that might affect the community, particularly concerning accidental or routine releases of hazardous materials from the chemical plant. These issues include operation of the on-site hazardous waste incinerator.

Chevron did not grant an on-site interview to INFORM but provided information through several telephone interviews. The plant reported a total of five source reduction activities reducing 140,000 pounds of waste each year. The company did not grant an interview for INFORM's 1985 study.

Products and Operations

This Chevron Chemical facility has made several changes in its product line since it was first profiled in INFORM's 1985 study. The manufacture of a fungicide has ceased because it had reached the end of its competitive product life and would have required registration — a process that Chevron considered too expensive to justify continued production. Also, consumer products such as fertilizers, pesticides, and herbicides used to be formulated and packaged at the Richmond site, but were moved to Chevron's midwest operations for economic reasons. New products manufactured at the Richmond site include gasoline additives and, at small-scale levels, an herbicide. These production and operational changes at the plant have resulted in an overall decrease in the number of contract and staff employees from 455 in 1983 to 274 in 1991.

Environmental Policy

The goals of the Chevron Chemical Company's written environmental policy are "to minimize the generation of industrial waste, both hazardous and non-hazardous to the extent economically and technologically feasible, and to handle industrial waste that is generated in a manner that minimizes future company economic and environmental liability." The policy identifies source reduction as a first priority, followed by recycling and treatment on-site or at company facilities, and finally recycling and treatment by "reputable" facilities that are in legal compliance with environmental regulations. The company audits all of these facilities prior to their use. Land or any subsurface disposal is used as a last resort. This policy went into effect in 1985-1986.

Chevron noted that California law SB14, the Hazardous Waste Source Reduction and Management Review Act of 1989, has been a major factor motivating it to adopt a source reduction policy and program. It requires all facilities producing at least 12,000 kilograms (13.2 tons) of federally regulated or California-designated hazardous wastes to prepare a written source reduction plan.

As a member of the Chemical Manufacturers Association (CMA), this Chevron plant is required to adhere to the CMA's "Responsible Care" program guidelines that require that "each member company shall have a waste and release reduction program... giving preference to source reduction." The CMA Responsible Care guidelines became mandatory for all CMA members in mid-1990.

Also, Chevron's corporate program, Save Money and Reduce Toxics (SMART), went into effect in 1986-1987. It requires Chevron facilities to contribute to a company-wide goal of 50 to 60 percent reduction of solid and liquid wastes by 1992. The plant's environmental health and safety manager reports progress annually to Chevron Chemical headquarters. Chevron Chemical then reports to the corporate Health, Environment, and Loss Prevention Program office. The report includes a 5-year plan for reduction programs in place, future plans for such programs, dollars spent, and quantity of emissions reduced.

Materials Data Collection

For over 20 years, Chevron has been collecting materials balance data at the process level to track yields per pound of raw materials used on a weekly and monthly basis. Processes are scrutinized for possible efficiency improvements. Chevron was motivated to establish this data collection system mainly because of the rising cost of raw materials and products that this facility produces and competition particularly from overseas companies.

All operations at the Richmond facility are subject to statistical process controls (SPCs). SPCs can reveal, among other things, sources of waste that might be generated because of improper or fluctuating process conditions; inadequate, old, or poorly maintained equipment; and operator inconsistencies. Everybody at the plant goes through a 2-day training class, and is given refresher classes when needed.

Furthermore, 80 percent of the processes at the plant are computer controlled, in order to ensure that optimum process conditions, identified through SPC data collection, become consistent over time.

Nonproduction areas of the plant, such as loading and unloading operations, are also included in the tracking process. Tracking begins with the bill of lading (the invoice of materials on trucks) that arrives with a tank truck of raw material, and ends with the weighed amount of product in a tank truck before shipping from the facility. The loading and unloading systems are enclosed and permanent; that is, they do not require cleaning between materials transfers.

Costs associated with waste generation at Chevron are allocated to the process from which the waste was generated. Each process at the plant is treated as an individual business. Costs of wastes destined for the on-site incinerator (mostly aqueous wastes containing 80 to 85 percent water, and point source air emissions from processes), RCRA and California List waste sent off-site, and fugitive air emissions are allocated back to the processes. Raw materials costs were already allocated to each process by the nature of the plant's organizational structure. Nonhazardous wastes such as floor sweepings are also allocated back to the source of generation.

Other Source Reduction Program Features

Responsibility for progress towards the corporate reduction goals and the internal plant goals lies with the environmental health and safety manager at the Richmond facility and is part of his performance evaluation. Also, managers of each facility within the Richmond site are accountable for source reduction progress. Authority for implementing individual source reduction activities lies at various levels of management depending on project size. This ranges from the shift supervisor and mid-management people, to the management teams and plant manager.

Since 1989, the Richmond facility has followed the Deming philosophy of management which calls for interdisciplinary teams to investigate process operations at the plant. The plant manager lists the objectives for each team. "Plant CAER Teams" are made up of people from operations and maintenance, including environmental, design, and process engineers. (CAER is an acronym for the Chemical Awareness and Emergency Response Program of CMA.) "Support Services Teams," made up of people from environmental health and safety, engineering, purchasing, and maintenance offer support for the Plant CAER Teams and work to ensure smooth interaction between the teams and those outside the plant. Ideas for yield improvements are solicited by the teams from operations personnel.

Source Reduction Activities

Chevron reported five source reduction activities to INFORM for this study; they are summarized in Table II-6 and described below. Chevron did not report any source reduction activities for INFORM's 1985 *Cutting Chemical Wastes*.

The first activities occurred over a 3- to 4-year period for a process that operates 7 days a week, 24 hours a day. Chevron reduced the amount of methylene chloride fugitive air emissions reported annually to TRI from 182,000 pounds to about 40,000 pounds in 1990. Chevron told INFORM that these emissions quantities are actual measurements and are consistent throughout the time period reported. This 78 percent reduction was accomplished through a variety of small process and operational changes resulting from three lists of employee suggestions and an equipment change. Double barrier seal pumps were installed which can detect leaks and allow operators to switch to other pumps while leaks are being repaired. The investment in new pumps was on the order of several hundred thousand dollars. Chevron told INFORM that the major factor motivating this change was that methylene chloride has been reviewed as an air contaminant since the early 1980s by California's environmental regulatory agency.

The second source reduction activity at the Richmond facility reported to INFORM resulted in a 40 percent reduction over a 15-year period of an aqueous waste stream (80 to 85 percent water) that used to go to the on-site incinerator. The project began in the early 1970s. Waste reduction and product yield improvement were accomplished through chemical substitution and operational changes; there was no change in final product formulation. While the original two chemicals and their substitutes are all considered hazardous, the substitution allows Chevron to use less raw materials to produce the same amount of product. The operational changes were employee suggestions.

The third source reduction activity resulted in a 50 percent decrease in the overall quantity of an aqueous waste stream (80 to 85 percent water) destined for the on-site incinerator. This source reduction activity, implemented in 1989, was the result of employee suggestions and involved a chemical substitution and a process change. As in the above case, while the original chemical and its substitutes are all considered hazardous, the substitution allows Chevron to use less raw material to produce the same amount of product. While information on the total dollars spent was not available to INFORM, Chevron reported that this source reduction activity paid for itself in less than 6 months.

The final two source reduction activities reported to INFORM involve quality assurance sampling and management procedures. Quality assurance sampling equipment is now automated so that only a single vial of sample is delivered. Samples are returned to the process whenever feasible, or are sent to the on-site incinerator. Similarly, all finished product samples are returned to the process; this last procedure has been followed for at least the last 20 years at the Richmond plant.

Table II-6   Chevron Chemical Company (Fremont, CA): Source Reduction Activities

Waste Medium (SR Type) Year	Source Reduction Activity	Specific Waste Reduced (Hazardous or Nonhazardous)	Percent Waste Reduced	Amount Waste Reduced
Air (EQ, PS, OP) 1987-1990	A variety of small operational changes and process modifications reduced fugitive emissions of dichloromethane. Also, double-barrier seal pumps were added to detect leaks and allow switch-over to other pumps while leaks are repaired.	Dichloromethane (H)	78%	140,000 lb/yr
Water (CH, OP) 1970s (over 15-yr period)	Two chemical substitutions allow less raw material to be used per pound of product produced.	(H)	40%	
Water (CH, PS) 1989		(H)	50%	
(EQ)	Automated quality assurance sampling equipment ensures that only a single vial of sample is delivered.	(H)		
(OP) pre-1970s	Some quality assurance samples and all product quality samples are returned to the process.	(H)		

Key to source reduction types: CH, chemical substitution; EQ, equipment change; OP, operational change; PR, product change; PS, process change.
A blank indicates that the plant did not provide information.

Future

The environmental health and safety manager at Chevron's Richmond plant told INFORM that the plant has prepared extensive plans for future reduction because such plans are required as part of its RCRA permit for the on-site incinerator and storage facilities, its membership in CMA, the California Hazardous Waste Source Reduction and Management Review Act of 1989, and the corporation's own SMART program.

He reported that "we're looking at practically everything": some of the plant's future source reduction activities will include upgrading pneumatic electronics to computer controls, improving raw material feed systems to processes, installing more accurate metering equipment, and replacing conventionally sealed pumps with closed magnetic induction pumps.

Change in Yield	Dollars Saved	Dollars Spent	Motivation	Comments	Time Needed for Implementation
		$200,000	Employee suggestion. Dichloromethane has been reviewed as an air contaminant since the early 1980s.	Dollars spent is an estimate for new pumps only.	
			Employee suggestion.	The original two chemicals and their substitutes are all considered hazardous.	
			Employee suggestion.	The original chemical and its substitutes are all considered hazardous. This source reduction activity paid for itself in less than 6 months.	

CIBA-GEIGY CORPORATION
Toms River, New Jersey

Summary

The large complex of Ciba-Geigy's Toms River plant in southern New Jersey opened in 1952; in 1987, two divisions, the Dyestuff and Chemical Division and the Plastics and Additives Division, were producing about 450 chemical products (dyes, epoxy resins, and additives). Following several years of conflict with New Jersey regulators and the local community over the plant's waste management practices, Ciba-Geigy began to phase out its operations in the mid-1980s. The Toms River plant closed its dye production facilities by the end of 1988 and has now also ceased all plastics and resins operations; it operates as an environmental testing and dye warehousing and packaging facility. The company's plans for constructing a pharmaceuticals facility at Toms River were stopped by the denial of a construction permit by the New Jersey Department of Environmental Protection. The employment has decreased from over 1,000 in 1983 to about 400 with the closing of the dye, plastics, and additives operations.

In 1984, a leak in the Toms River plant's 10-mile steel pipeline that discharged wastewater into the Atlantic Ocean incited several media-attracting actions by Greenpeace. Ciba-Geigy repaired the leak, but the incident touched off a prolonged period of negotiations between the local community, the plant, and EPA. The issues were the continued existence of the pipeline and clean-up of the Toms River site, which had been put on Superfund's National Priority List.

A final court-approved settlement between Ciba-Geigy and the state of New Jersey was reached in February, 1992. The company agreed to pay $50 million, or more if needed, to clean up the Toms River site and a polluted aquifer, as well as smaller amounts in civil and criminal penalties and administrative costs. It will also make a donation to help the state purchase wetlands. The ocean discharge pipe was closed at the end of 1991.

While manufacturing operations at the Toms River site have ceased, the plant still provides examples of effective source reduction activities. Further, it had a comprehensive source reduction program containing all of the features identified by INFORM. Finally, at the corporate level, Ciba-Geigy has sought to institutionalize source reduction; at the Toms River Plant, for example, plant officials looked for ways to incorporate source reduction ideas when planning new processes and capital projects. Thus, this profile discusses the source reduction program features in place when manufacturing operations were ongoing and describes the source reduction activities implemented at the plant before manufacturing ceased.

In 1985, in order to obtain a renewal of its ocean discharge permit, Ciba-Geigy undertook an extensive, plant-wide toxicity reduction study of all wastestreams leading to its on-site treatment plant. This study not only resulted in many source reduction projects but also prompted the implementation of a systematic quality improvement program that featured source reduction as a key element and included a materials balance and full cost accounting system to track all types of waste and incentives for reducing waste. The corporate environmental policy assigned specific responsibilities to each employee and included guidelines for assessing source reduction during research and development.

Ciba-Geigy granted an on-site interview to INFORM in 1987 and conducted a tour of its Toms River, New Jersey, facility for this new research. The plant reported a total of 11 more source reduction activities, reducing 293,000 pounds of waste and saving the company $1,593,100 each year. Ciba-Geigy had also granted an interview for INFORM's 1985 study and at that time reported five other source reduction activities.

Products and Operations

At present, the Toms River plant is operating only as an environmental testing and dye warehousing and packaging facility. However, up until the fall of 1988, the products made by Ciba-Geigy's Dyestuff and Chemical Division and Plastic and Additives Division at the Toms

River plant included about 450 dyes, epoxy resins, and additives. The dyes were produced for use by the textile, carpet, paper, leather, and automotive industries. After ceasing dyestuff production, the plant continued to manufacture over 100 different epoxy resins, although this has now stopped as well. These resins and adhesives were used in the aerospace and construction industries to provide lightweight construction materials and structural integrity of products ranging from bridges to the space shuttle. Speciality coatings were used by the marine and construction industries to provide protection against corrosion and the elements. Additives were used by the automotive and other industries to provide resistance to heat, aging, and fading from sunlight. Other additives improved the performance of lubricants used by the electric power, automotive, and aviation industries.

Reduction in production occurred at this plant in recent years in the face of years of conflict between EPA, New Jersey environmental authorities, the local community, and Ciba-Geigy. In 1983, the Plastics and Additives Division at Toms River discontinued production of high-volume commodity resin, replacing it by an increase in production of speciality resins. This resulted in a 20 percent increase in the number of products and a 10 percent decrease in product volume. In 1985, three dye production buildings were closed, resulting in a 30 percent decrease in product volume for dyes and a 50 percent reduction of intermediates produced. In 1987, a fourth dye production building was closed, causing a further reduction of intermediate production by 85 percent and of dye production by 15 percent from its 1986 level. All dye production ceased on September 30, 1988 when the last production building was closed; most dye production has been transferred to Ciba-Geigy's plant in St. Gabriel, Louisiana.

The 1986 corporate reconstructuring plan called for other changes at the Toms River plant, including phasing out all plastics and resins operations by the end of 1990. Continuing dye operations consist of formulating dyes only, using imported dyes, and blending and packaging them according to customer specifications.

Ciba-Geigy's restructuring plans had called for the construction and operation of a $90 million pharmaceuticals manufacturing facility at Toms River by 1992. However, in October, 1988, the New Jersey Department of Environmental Protection denied a construction permit for the new facility. The corporation decided not to appeal the state's ruling and to continue to import products while it reassessed its plans and manufacturing needs.

A corporate Environmental Testing Laboratory has been established at the Toms River site. This laboratory conducts environmental sampling and testing, such as groundwater monitoring, for this and other Ciba-Geigy plants. The company deemed this corporate support function necessary because it experienced difficulty getting timely and high quality work from commercial laboratories.

In 1987, the number of employees at the Toms River plant was about 900, down from over 1,000 in 1983. With the closing of the dye, plastics, and additives operations, employment is now about 400.

Environmental Policy

The formal written environmental policy of the Ciba-Geigy corporation states that all manufacturing operations are to be carried out in a way that will not adversely affect the environment and that all operations are to be in compliance with applicable laws and regulations. This environmental policy states that proper disposal of all wastes is required as part of the production process.

While not part of the formal written environmental policy, the corporate policy for waste management, as reported to INFORM, is to manage wastes according to the following hierarchy of waste management: (1) reducing the wastestream at source wherever possible; (2) if this is not possible, recycling the waste; and (3) if recycling is not possible, using high-temperature incineration. Land disposal, in properly designed and permitted facilities, is used only as a last resort, and then preferably on company property where it can be monitored and controlled.

Ciba-Geigy's corporate environmental policy assigns specific responsibilities to each employee. The plant manager bears overall responsibility for the proper conduct and performance of the plant on environmental matters. Each plant must also have an environmental manager who acts as a staff expert on environmental problems; assesses the environmental aspects of operations; maintains and updates files, records, and permits; and organizes and coordinates environmental audits. The corporate Office of Environmental Protection and Services acts as an advisor and consultant to the individual plants' managements.

The corporate environmental policy also contains guidelines on how process research and development is to be carried out. The steps specified include environmental problem assessment during the research and development phase and a materials balance for each processing step. All existing, new, revised, or transferred processes fall within the scope of this policy. The environmental assessment covers air emissions, liquid wastes, and solid wastes, as well as ecotoxicological properties of raw materials, intermediates, and end-products. Emphasis is placed on proper disposal of all wastes as part of the production process.

Corporate policy also sets the requirements for environmental audits focusing on regulatory compliance. A permanent group of professionals, including chemical engineers and analytical chemists familiar with the production processes, performs annual environmental reviews at each Ciba-Geigy site. The focus of the review is on compliance with local, state, and federal regulations and internal Ciba-Geigy guidelines. Because the same team conducts these reviews each year, they are familiar with the processes and regulations involved. Each year, the reviews are more detailed.

Materials Data Collection

Each process at Ciba-Geigy's Toms River plant had a materials balance done for it, and wastes were tracked by product. Wastes tracked in the materials balance included all solid wastes, the content of the aqueous wastestreams, and air emissions. Wastewater generated by the various plant areas was first tracked in 1986. The accuracy of the material balance was within 5 percent of the actual measured wastewater generation.

The data gathered were used in developing budgets that included targets for reducing wastes discharged to air, water, and land. The plant environmental manager conducted quarterly plant reviews of progress in developing and implementing source reduction projects and submitted reports that were circulated throughout the corporation.

Full cost accounting was accomplished by charging waste disposal costs incurred by each process to that process. For wastewater streams, each building at the Toms River plant site was charged per unit of flow, BOD (biological oxygen demand), and TOC (total organic carbon). For solid wastes, charges were based on handling and analytical services for specific chemicals or waste categories. Disposal costs were often high: for example, at the Toms River plant, in 1987, the budgeted cost to incinerate a 55-gallon drum was $907 and to landfill it was $297. In some cases, the incineration budget cost was more than the material purchase price. Continuous monitoring and sampling were used to establish the charges; the data were also used to verify materials balance calculations. Production managers were charged for actual wastes generated, with performance tied to these costs. Besides disposal and treatment costs, budgeted costs charged to processes included expenses associated with compliance (record keeping, storage, handling, and analytical services), insurance, accidents and waste clean-up, and public and customer relations dealing with waste issues.

Ciba-Geigy is establishing a corporate-wide database on wastes that will track nonhazardous and hazardous solid wastes, wastewater, and air emissions by concentration and/or weight. It was scheduled for full implementation by 1992.

Other Source Reduction Program Features

In addition to having a written source reduction policy, materials accounting/materials balance, and full cost accounting, Ciba-Geigy's Toms River plant also had the other four source reduction program features tracked by INFORM: leadership, an environmental program, environmental goals, and employee involvement.

In 1984, the Toms River plant launched a comprehensive program, called the Quality Improvement Process (QIP), to address all aspects of its operations, including source reduction. This process is used to train and motivate employees to do the job correctly the first time. Employee training exists for all operations and individuals. In addition to training, employees are given tools designed to ensure that they follow the company's guidelines, including measurement charts, error cause removal, and corrective action processes. Motivation is provided through a formal recognition process for employees who consistently achieve defect-free results.

Problem solving under the QIP generally calls for the creation of a Corrective Action Team (CAT). The systematic QIP approach used by such a task force consists of the following steps:

1. Define the problem
2. Identify the root cause
3. Fix the problem
4. Implement the corrective action
5. Evaluate and follow up
6. Recognize the efforts of the task force

Officials at Ciba-Geigy explained that while quality as a concept has always been a driving force at Ciba-Geigy, the goal is to prove the customer with the highest quality product at the best possible price. In the 1950s and 1960s, the quality concept meant zero defects and concentrated on process yield improvement and operator training. When waste disposal costs were a less significant part of total costs, the quality criteria were mainly applied to manufacturing operations. However, as disposal costs have increased (they can be as much as 50 percent of the operating costs) and environmental regulations have become more stringent, the quality criteria have been extended to all aspects of the plant's operations.

In the 1970s, motivated by the need to build new wastewater treatment and environmental protection systems, the corporation undertook a comprehensive review of source reduction opportunities. All of its manufacturing units were required to review their operations and to develop and implement short- and long-term action programs.

The formal corporate-wide effort to reduce in-process waste began in 1984. Development personnel, together with production and division support, are required to prepare material balances for processes, define sources of waste, and work to reduce these. Bimonthly progress report meetings are scheduled with the plants. The corporation also holds quarterly production coordination meetings dealing with environmental issues, including reduction of waste.

Ciba-Geigy officials state they have taken three approaches to source reduction. The first and quickest is to exert tighter operating controls, especially in batch operations producing many different products. The second is to improve the process, which can take anywhere from 6 months to 2 years. A longer time frame is involved for the third approach: adopting fundamentally new processes or new technology, including the design and construction of new facilities. The company reports that its greatest successes have been found in process improvements and changes.

Company officials also reported that personnel from both production and marketing have been involved in the search for source reduction opportunities because source reduction affects all aspects of the product, particularly product quality. For example, cleaning out kettles produces waste but, without cleaning, cross-contamination may result. Recycled raw materials can result in additional waste if they contain impurities that might cause off-specification product to be produced. In addition, substitution of raw materials may affect customer satisfaction and product quality.

Ciba-Geigy is currently developing a computer network to track waste management needs, requirements, and compliance at all its sites. The system, known as the Environmental Data Management System, was scheduled to be fully operational by 1992.

At Toms River, Ciba-Geigy dedicated plant management and plant personnel and significant resources to studying its wastewater problems. It developed and prioritized process wastestream profiles and set reduction goals and strategies for meeting them. It rewarded people for exceptional performance by giving them Quality Improvement Process awards. More than 10 people at Toms River received these awards in 1987.

In addition, the Toms River plant, in 1984, instituted a toxicity reduction program to meet the increased requirements expected in its upgraded ocean discharge permit, which was renewed for 5 years in 1985. The Quality Improvement Program, as applied to source and toxicity reduction at Toms River, involved first creating a task force consisting of chemists, chemical and environmental engineers, and maintenance and operating personnel to study the problem and develop a plan for corrective action. Results of the implementation of the plan are described below.

Source Reduction Activities

Ciba-Geigy reported implementing 11 source reduction activities since INFORM's 1985 report, *Cutting Chemical Wastes*; these are summarized in Table II-7 and described below. The company had reported five other source reduction activities that were described in the earlier report.

Ciba-Geigy has documented annual cost savings of $1.6 million at the Toms River plant, between 1984 and 1987, by calculating the first year of savings from each source reduction effort. These efforts were accomplished with less than $300,000 in equipment costs.

Dyestuff and Chemical Division. The need to meet new standards imposed by New Jersey in 1985 in renewing Ciba-Geigy's ocean discharge permit motivated four of the five source reduction activities reported by the Dyestuff and Chemical Division at the Toms River facility. The plant had a pipe that discharged treated wastewater from its on-site wastewater treatment plant to the Atlantic Ocean 10 miles away. The new permit mandated reducing the toxicity of this discharge wastewater (including biological toxicity as well as heavy metal content).

Toxicity reduction involved a two-pronged approach: (1) reducing the toxicity of the wastestream sent to the treatment plant (called influent), while increasing its biodegradability, and (2) improving wastewater treatment technology. The first step was to educate plant personnel and management about the need for toxicity reduction. Ciba-Geigy officials stated that a successful Quality Improvement Program (QIP) requires commitment and support from top management from the outset. Task forces, organized by each plant production unit (since all production units generated wastewater streams), were assigned to identify, evaluate, assign priorities for, and implement source reduction opportunities for their production line.

To characterize the waste treatment plant influent, materials balance calculations of all the production processes were used. The goal, which was exceeded, was to sample and analyze at least 90 percent of the process wastewater streams. The samples were tested by the plant's environmental testing laboratory for total organic carbon (TOC), biological oxygen demand (BOD), total suspended solids (TSS), acidity (pH), biodegradability, and toxicity. The acute bioassay toxicity tests required as part of this study used mysid shrimp, one of the sensitive test species, and cost more than $1 million. Out of more than 500 wastestreams, 17 were classified as toxic and nonbiodegradable and 29 as toxic and biodegradable. Source reduction efforts then concentrated on these 46 wastestreams.

An environmental task force, created within the research and development group, identified process changes and improvements throughout the plant that would enable the wastewater to meet the new discharge limitations. This task force, composed of six chemists and six senior laboratory technicians, concentrated on the filtrates from 29 products that exhibited severe toxicity to mysid shrimp. Interim measures to alleviate the problem were devised while permanent solutions were being developed. Action plans were written with short-term solutions in conjunction with longer-term research programs. For many wastestreams, the short-term solution was treatment at the source.

One focus of attention was heavy metals since the list of product wastestreams showed that products whose production involved a heavy metal at some stage of the synthesis usually resulted in filtrates with relatively high toxicity. The plant used heavy metals (specifically, copper, chromium, and zinc) in dyestuff manufacture for the formation of metal complex dyes (and, in some reactions, as catalysts). These dyes do not fade and, where customers require this feature (for example, in automotive fabrics), these dyes could not be replaced by products that do not contain metals. Many previous attempts had been made to reduce the discharge of metals from the processes used to make these dyes. For example, the previous INFORM report referred to a new solvent chroming procedure. However, this method failed to replace the existing aqueous process because of the problems associated with operating a solvent system (toxicity of the solvent, equipment for solvent recovery, air emissions, disposal of still bottoms, fire risk, etc.).

The short-term solution in this case was to remove the heavy metals from all filtrates by precipitation at the process source, before mixing wastestreams. Such precipitation reduced heavy metals in the wastewater stream by over 90 percent, to less than 5 pounds per day. To reduce the volume of solid residues, a special (J-mate) dryer was installed. This equipment reduced by two-thirds the volume of solid metallic residues sent to off-site RCRA-permitted landfills. Ciba-Geigy recognized that this precipitation was not source reduction since it just transferred the heavy metals from aqueous effluent to a solid residue. Hence, longer-term process improvements were explored.

For the first source reduction activity reported to INFORM, two types of process improvements were used. In some cases, excess metal could be controlled to very tight limits by the use of modern analytical methods. In other cases, it was possible to dry the whole reaction mass and thus eliminate the discharge of metal entirely by shifting the necessary purification steps to an earlier part of the process. Starting in 1985, the Toms River plant made six products using these methods. They resulted in an increase in yield of 11 percent, complete elimination of heavy metal waste from these processes, disposal of 100 fewer drums (45,000 pounds) of waste per year, and annual cost savings of $86,500.

Another process change, initiated in 1986, involved working with purified intermediates so the metallization filtrate could be recycled, at least for a number of batches when a substantial excess of metal was necessary to complete the metallization reaction. Preliminary indications showed that costs could be reduced by these methods because yields at the final step were about 10 percent higher and precipitation costs for removal of the metals were avoided. Overall annual costs savings were $11,600, with disposal of 40 fewer drums (18,000 pounds), representing an 80 percent reduction in wastes.

The action plans also identified two changes in the multistep dye-making process (a chemical substitution and process improvements) that made possible a 40 percent increase in yield, reduced iron waste by 100 percent and total organic carbon waste by 80 percent for the process, and resulted in annual cost savings of $740,000. The first change took place in the final step of the dye-manufacturing process where an aromatic nitro compound was converted to an amine. This step was formerly carried out with iron in a Bechamp reaction but, because of the large amount of solid iron sludge that formed, iron was replaced with a different conversion reagent. The second change involved reducing the presence of amine product in the filtrate, which had been shown to be the cause of the high toxicity of the effluent from this reaction to mysid shrimp. Improving the process during the conversion step eliminated the loss of product and significantly reduced the toxicity of the filtrate. Initiated in 1986, these combined source reduction actions reduced the generation of the amine waste by eliminating separation of the nitro compound which was required before the conversion step in the original process.

Additional development work on the first step of the synthesis showed that this yield could be further increased by 15 percent by altering the reaction conditions and reactants. This final improvement was not implemented in Toms River because of the transfer of manufacture to Europe. However, the new process has been transferred with the product.

Process improvements in the manufacture of a complex poly-azo dyestuff also paid off for Ciba-Geigy. The toxicity to shrimp of the filtrate of this product was shown to be due to *m*-phenylene diamine coupled to this dye in the final step of the manufacturing process. Each step of the synthesis was examined and the composition of the by-products determined. Major losses

Table II-7   Ciba-Geigy (Toms River, NJ): Source Reduction Activities

Waste Medium (SR Type) Year	Source Reduction Activity	Specific Waste Reduced (Hazardous or Nonhazardous)	Percent Waste Reduced	Amount Waste Reduced
Solid/water (PS) 1985	Excess metals eliminated in dye manufacture by shifting purification to early part of process and improving process controls.	Heavy metals (H)	100% for six products	100 drums (45,000 lb/yr)
Solid/water (PS) 1986	Use purified intermediates so can reuse metallization filtrate when an excess of metal is needed to complete the reaction.	Heavy metals (H)	80%	40 drums (18,000 lb/yr)
Solid/water (PS, CH) 1986	Process change to eliminate nitro separation step and replace iron as raw material.	Iron sludge (H)	100% of iron and 80% of total organic carbon (TOC)	
Water (PS) 1986	Process change to reduce amount of *m*-phenylene diamine used and discharged in filtrate water.	Amino coupling component (H)	80% of TOC in filtrate	
Solid (OP) 1985	Use minimum quality control sample sizes and return samples to process.	Intermediate (N)	50%	20,000 lb/yr
Solid (PS) 1987	A new antioxidant process eliminates the need for excess hydrazine.	Hydrazine precipitate (H)	100%	90,000 lb/yr
Solid (EQ) 1987	Installed separate dust collectors in new powder coatings process. Separates dust emissions so can be reused or sold as product.	Dust (N)		50 lb/batch (20,000 lb/yr)
Air (EQ) 1985	Designed sampling and charging devices for kettles in the resin solution production process to reduce solvent emissions.	Solvents (H)	90%	50 tons/yr (100,000 lb/yr)
Air (PS,EQ)	Use pumps to transfer liquids rather than blowing with nitrogen.			
Air (PS)	Installed computer control for maintenance of inert nitrogen atmosphere.			
Air (OP)	Limit open manhole operations.			

Key to source reduction types: CH, chemical substitution; EQ, equipment change; OP, operational change; PR, product change; PS, process change.
A blank indicates that the plant did not provide information.

Change in Yield	Dollars Saved	Dollars Spent	Motivation	Comments	Time Needed for Implementation
+11%	$86,500	$0	New ocean discharge permit standards and high cost of disposal.		2 mo
+10%	$11,600	$0	Eliminate toxic components to meet new ocean discharge permit standards.		1 mo
+40%	$740,000	$0	Eliminate toxic components to meet new ocean discharge permit standards.	Reduced loss of product in wastewater and need for filtration.	9 mo
+12.5%	$250,000	$0	Eliminate toxic components to meet new ocean discharge permit standards.	Five percent reduction in use of *m*-phenylene diamine in the process.	3 mo
	$100,000	$0	Disposal cost savings.	Expense of handling samples to reduce risk of cross-contamination is justified by rising disposal costs.	
	$335,000	$200,000	Increased costs to recycle and recover raw material.	Recovery had been done by supplier that did not want to register as a treatment, storage, and disposal (TSD) facility.	2 yr
+0.2%	$20,000/yr	$80,000	Cost of product and disposal.	This process underwent the source reduction analysis required for all new products at Ciba-Geigy.	
+1%	$50,000/yr	$10,000	Regulatory compliance.		1 mo
				Minimize nitrogen flow and, thus, entrainment of volatile materials.	

were found to occur during the second diazotization step. By changing conditions at this step, starting in 1986, overall yield was improved by 10 to 15 percent, 5 percent less *m*-phenylene diamine was used, total organic carbon (TOC) of the filtrate was reduced by 80 percent (reducing the amount discharged to levels that were no longer toxic to shrimp), and annual cost savings were $250,000.

The fifth source reduction activity that the Dyestuff and Chemical Division reported implementing involved operational changes related to the handling of samples. Previously, product samples were analyzed and then disposed of, for fear that if a sample was returned to the wrong batch, the whole batch might become waste. Starting in 1985, the plant began taking the minimum sample size needed and returning the samples to the process. The expense of having a staff chemist oversee the handling of samples to avoid cross-contamination became justified as costs of waste disposal increased. Returning samples to the batches reduced wastes by 20,000 pounds (a 50 percent reduction) and saved the company $100,000 per year.

Although not reported as a specific source reduction activity, other operational control changes also led to significant source reduction and yield increases. In the manufacture of dyestuffs, the final separation of the product is often a purification step. It is necessary to balance a loss of some dyestuff in the filtrate, along with impurities, against the quality of the product. Largely because of new levels of technology and analytical capability developed in recent years, improvements in process controls made it possible to reduce the amounts of by-products formed during the multistage processing that dyestuffs require.

Before the decision to cease production of dyes at Toms River, Ciba-Geigy was developing another way of reducing waste from synthetic dye-making. Traditionally, the reactions used to make synthetic dyes take place in large reactors, from 3,000 to 7,000 gallons in size. One of the problems in using these large reactors has been the length of time needed to complete the reaction, which resulted in product decomposition. The new technology which Ciba-Geigy's Toms River plant would have applied to its dyestuffs production included small or more efficient multistage reactors controlled by computerized control systems to optimize the reaction cycle. Precise control and short cycle times result in less decomposition and, hence, less waste. An estimated 10 to 15 percent increase in yield was projected. The TOC reduction in the wastewaters was expected to be two-thirds had this new technology been implemented.

Plastics and Additives Division. The Plastics and Additives Division at the Toms River plant reported on the implementation of six source reduction measures, two involving solid wastes and four involving air emissions. In one, in 1987, the Toms River plant implemented a process change that cost $200,000 to implement, reduced 90,000 pounds of solid waste a year (completely eliminating a hazardous hydrazine precipitate), and resulted in annual cost savings of $335,000. Previously, the process for production of an antioxidant had required use of an excess of hydrazine, which then had to be precipitated before solvent recovery. Originally, the precipitate was disposed of by incineration or landfill. Then, for a short time the precipitate was recycled to the supplier for recovery. But, when government regulations required the supplier to obtain a treatment, storage and disposal (TSD) facility permit to continue recycling, the supplier did not want to go to the time and expense of obtaining the permit and stopped recycling the precipitate; Ciba-Geigy returned to incinerating it. In order to eliminate the hydrazine precipitate and thus the need to incinerate it, Ciba-Geigy's Process Development Department spent 2 years to develop new chemistry to produce the antioxidant.

The other change reducing solid waste resulted from Ciba-Geigy's requirement that source reduction be addressed whenever a capital project or process change is developed. In this case, a different granulator was needed in order to provide a smaller, more uniform resin flake for powder coating customers. Powder coatings, used to coat appliances, metal furniture, and automobiles, have the advantage that they are paints that do not use solvents. Instead, the coating particles are electrostatically sprayed onto the metal part. The part is heated in an oven where the powder particles melt, flow, and form a smooth coating. In providing a smaller flake size, the new granulator was expected to produce an additional 50 pounds of dust per batch. As a result of considering source reduction during process development, separate dust collectors were installed in order to keep the various dusts separate so they could be reused in the next batch or sold as a product. The cost of the equipment was $80,000. The project was implemented during 1987 and resulted in an increase in yield of 0.2 percent and savings in the first year of operation of $20,000.

Air emissions at Toms River were calculated for every product and process. These calculations were verified by actual measurement on some products. The emission points (pumps, vents, flanges, and valves) were identified and the emissions collected and analyzed using a gas chromatograph. Based on this work, an action plan was developed to reduce emissions of volatile organic substances (VOS). Since most losses of solvent were found to occur during sampling and charging procedures in resin solution production, the Engineering Department designed new sampling and charging devices for a number of kettles to reduce the emissions. These devices, installed in early 1986, reduced emissions by approximately 50 tons per year (90 percent) and improved yields by 1 percent. The cost of the devices was $10,000, and their use resulted in a savings of $50,000 per year.

The three other source reduction projects to reduce air emissions that the Plastics and Additives Division reported were: (1) using pumps to transfer liquids rather than blowing with nitrogen; (2) installing a computer control system for maintenance of inert nitrogen atmosphere to minimize nitrogen flow, thus minimizing entrainment of volatile materials; and (3) limiting open manhole operations.

Future projects were also planned. The company also reported developing additional equipment design changes to reduce emissions of volatile organic substances: (1) conservation vents and vapor balance lines on storage tanks; (2) vent condensers in addition to main reactor condensers to enhance recovery of volatile materials; and (3) closed-loop recirculation vacuum pumps, thus allowing recovery of condensed organics and reducing atmospheric emissions and losses of entrained organics to the process and to scrubber waters during vacuum operations.

Planning for the New Pharmaceutical Facility. Although Ciba-Geigy was denied a construction permit for a new pharmaceutical facility at the Toms River, the proposed facility contained several examples of process innovation directed towards reducing and/or eliminating priority pollutants in wastewater and potentially toxic pollutants. In one, an alternative to the FDA-approved process for synthesizing one of the active drug substances was being developed to ensure that chloroform would not end up in wastewater or as air emissions. The new process would have eliminated the use of trichloroacetic acid as a raw material because a by-product from this raw material in the current process is chloroform.

Ciba-Geigy also investigated the elimination of ammonia in the purification of one active ingredient and substitution of an alternative neutralization base for the ammonia in the chemical synthesis of another drug substance. The reduction or elimination of ammonia in the wastewaters was expected to reduce its toxicity.

In addition, aqueous process liquors from several of the active ingredient purification processes would have been recycled back to the previous manufacturing steps. As a result, the discharges of wastewater streams from these processes would have been reduced or eliminated. A storage tank for recycled liquors from one of these products was included in the project scope for the pharmaceuticals facility. A modification of distillation conditions for recovery of excess isopropyl amine has resulted in a 30 percent reduction of losses at another Ciba-Geigy facility and would have been used at this plant as well.

Other Waste Management Practices

In addition to its source reduction activities, Ciba-Geigy also reported a number of other waste management practices. For instance, in anticipation of the more stringent ocean discharge requirements imposed by the New Jersey Department of Environmental Protection in 1985, Ciba-Geigy undertook a comprehensive toxicity reduction program that led in some cases to replacing toxic with nontoxic production ingredients. But, where this was not possible, the company either discontinued production of products or added a pretreatment step in the production area prior to the plant wastewater reaching the wastewater treatment facility. Through laboratory and pilot-scale testing of several technologies, the addition of powdered activated carbon to the plant's biological treatment process emerged as the most effective way to further reduce the toxicity of the final treated wastewater.

Through these processes, the plant was able to achieve 95 to 100 percent survival rates for the mysid shrimp (the naturally occurring test species) in 100 percent treated wastewater. These results are better than the permit limits which required a final survival rate of 50 percent in a 50/50 percent mixture of seawater and treated wastewater. Company officials also noted that they were achieved 18 months ahead of the schedule required by the permit.

In issuing the 1985 ocean discharge permit, the New Jersey Department of Environmental Protection also required the Toms River plant to submit a monthly discharge monitoring report. This report showed that, in 1987, the total organics present in the Toms River plant's treated wastewater were 50 percent lower than they were in the second half of 1985 (equal to a reduction of 4 pounds of total organics per thousand pounds of products produced). The chemical constituents were also reduced substantially. Table II-8 shows the changes in biological oxygen demand, total suspended solids, nitrobenzene, and three metals in the wastewater effluent from the Toms River plant from 1979 to 1987.

Table II-8 Ciba-Geigy (Toms River, NJ): Wastewater Effluent, 1979-1987

Parameter	1987 (mg/l)	1986 (mg/l)	1979-1983 average (mg/l)
Biological oxygen demand	9	12	73
Total suspended solids	11	21	154
Chromium	0.023	0.05	0.29
Copper	0.059	0.0174	0.61
Zinc	0.016	0.04	0.32
Nitrobenzene	0.000	0.00093	0.29
	Mean 1987	**Mean 1986**	**Mean 1985**
Toxicity testing (% LC_{50} in final effluent)*	100%	75%	15%

* LC_{50}, the concentration at which 50 percent of the organisms die and 50 percent survive, is used as a measurement of toxicity. The measure shown here is the percent concentration of the effluent at which LC_{50} is achieved.

These wastewater reductions were first achieved by removing toxic chemicals (mainly heavy metals) from the wastewater and treating them as solid wastes. This increased plant solid wastes in 1986 by 10 percent over those in 1985. Since 1986, the New Jersey Department of Environmental Protection has required hazardous waste generators to submit an annual source reduction report. The 1987 report shows that waste generated at Toms River for off-site disposal was almost 60 percent less in weight than that generated in 1985.

Nickel removal from the wastewater effluent was targeted as an action item in 1984 because of concern over heavy metal toxicity and meeting the expected increased requirements of the wastewater discharge permit. Production of one additive created mother and wash liquors that contained approximately 25 pounds of nickel per batch. Laboratory work showed that precipitating a nickel salt from the liquors and removing it by filtration was feasible. A process change was instituted in 1985 to precipitate the nickel salt which was then returned to the supplier who processed it to recover the nickel. Ciba-Geigy's precipitation process removed more than 90 percent of the nickel (170,000 pounds per year) from the wastewater streams. The cost of the project was $30,000, and the cost savings were $230,000. Production of the additive was discontinued at the Toms River plant in 1987.

A combination of reducing the toxicity of the stream entering the treatment plant and improving the treatment technology was needed because, even after toxicity reduction of the process wastestreams (sent to the on-site treatment plant), the biotoxicity of the wastewater leaving the treatment plant did not meet the level of toxicity reduction desired. One dilemma was that the toxicity of the wastestream leaving the treatment plant did not seem to be directly related to the toxicity of the wastestream it received for treatment. This may have been due to the toxicity of high-molecular-weight metabolic by-products produced during the biological sewage treatment process itself. In the continuing search for the cause of the problem, Ciba-Geigy sponsored a research fellowship at Vanderbilt University (a leader in the research and development of biological wastewater treatment technologies) to study this phenomenon and to determine whether it was specific to Toms River or is universal.

Technical Assistance

Ciba-Geigy relies mainly on its own expertise worldwide for information on process improvements and other waste-reducing programs. As far as source reduction programs are concerned, Ciba-Geigy has found that although the concept is simple to understand, in practice there is a lot of effort required to actually make it work in a particular situation.

Ciba-Geigy stated that technical information from the federal EPA is available but is difficult to find. Company officials indicated that information useful for toxicity reduction exists, particularly in the EPA laboratories, but that they often found that EPA's regulators were not familiar with it, suggesting a need for better communication between the scientists and the regulators.

Company Comments on State and Federal Regulations

According to Ciba-Geigy, the overall impact of government actions on source reduction activities has been favorable. Government regulations, particularly the federal requirement that a plant certify that it has a source reduction plan, have brought source reduction to the attention of industrial plant managers. Within plants, there is increasing information on waste generation, due in part to new government reporting requirements, and, hence, more recognition of opportunities for source reduction.

The Toms River plant manager said that he rarely has objections to legislation passed based on a legitimate need, but that he had concerns about interpretations by the regulators. In a previous assignment as a plant manager for another company in Cranston, Rhode Island, he felt it important to push for a clause in enabling legislation that regulations developed must be made in consultation with affected parties and community representatives outside the government. He reported that this requirement has proven successful in Rhode Island in creating more workable regulations that take into account the experience of industries and other experts.

Ciba-Geigy cited long delays in the permitting process as a deterrent to source reduction because years may be required to get permit approvals for innovative changes that may require additional permits.

In addition to signaling a need for more rapid permit processing, the company suggested that government source reduction reports take account not only of reductions in existing processes but also of measures designed into new plants or processes to prevent waste generation. Many source reduction reports, including New Jersey's waste minimization report, focus on comparing the volume of waste produced from a given process in different years, ignoring accomplishments that result from new processes. For example, because the recycling of dust, described above, was included in the process design phase, dust was never generated that could then be reduced. Thus, this prevention of waste would not be identified as source reduction in the reports. However, had the granulator been installed first, and separate dust collectors installed subsequently, this would have appeared in the required reports as a good example of source reduction.

Future

Ciba-Geigy sees the major area challenging federal and state government as that of reducing air emissions. Ciba-Geigy officials stated that Title III of the Superfund Amendments and Reauthorization Act (SARA) will bring about a significant reduction of industrial air emissions and that, for the chemical industry, these emissions will be better quantified and managed.

COLLOIDS OF CALIFORNIA
Richmond, California

Summary

The Colloids plant in Richmond, California, established in 1967, is a very small plant (four employees) located in Contra Costa County, north of San Francisco. It blends and compounds various liquid and dry materials for use as industrial anti-foam agents. It uses no regulated hazardous chemicals. Since 1986, when it was bought by Interchem of Louisville, Kentucky (Interchem was subsequently bought by Rhône-Poulenc), it has been part of a large chemical corporation and thus has had access to more information about environmental regulations. Its primary emphasis remains, however, on good housekeeping practices to keep waste to a minimum.

While Colloids did not grant an interview for INFORM's 1985 study, the company granted an on-site interview to INFORM for this update and conducted a tour of its Richmond, California facility. The plant reported a total of three source reduction activities.

Products and Operations

The Colloids plant does simple blending and compounding of various liquid and dry materials for use as industrial anti-foam agents. It also repackages additives and polyacrylates. No regulated hazardous chemicals are used at this plant.

Two tanks are used for blending. They are not rinsed between batches 99 percent of the time because the products are compatible. Colloids has installed a blender for powder anti-foamers because more of its customers now prefer dry products to liquid ones.

The plant is one of four plants owned by Colloids, Inc., with headquarters in Newark, New Jersey. In New Jersey, at the corporate headquarters, there is a regulatory affairs staff upon which the manager of the California plant depends for environmental, health, and safety information.

Environmental Policy

Colloids, Inc. has no formal written environmental or source reduction policy. The only impact on the plant from its purchase by Interchem was more reports on production, including an estimate of losses, and another on-site visit in addition to the annual one by Colloids. Both Colloids and Interchem perform annual on-site reviews, and there are memoranda from both informing the Richmond plant manager of new or revised procedures for environmental, health, and safety activities. The change in ownership when Rhône-Poulenc bought Interchem has brought about some new directives and new training manuals but has not affected the operations of this plant.

Materials Data Collection

No specific data other than production amounts are collected at Colloids.

Other Source Reduction Program Features

There is no official source reduction program at this small plant. However, the plant manager stressed that good housekeeping practices at this plant are strictly applied, in part because of his previous experience in pharmaceuticals manufacturing. He worked at a pharmaceuticals laboratory before coming to Colloids, and reports that he became well versed in worker health and safety. As part of the right-to-know procedures at this plant, there are labels that specify material safety data on all packages. New customers are also sent a material safety data sheet showing what the codes mean.

Table II-9 Colloids of California (Richmond, CA): Source Reduction Activities

Waste Medium (SR Type) Year	Source Reduction Activity	Specific Waste Reduced (Hazardous or Nonhazardous)	Percent Waste Reduced	Amount Waste Reduced
Solid (EQ)	Installed a dust collector on a new dry blending tank. At the end of the run, collected material is emptied and put into a drum and sold.			
Water/solid (OP)	Return samples of each product batch, as well as quality control samples, to the product batch.			
Solid (OP)	Use reusable Tote bins to supply product for one customer.			

Key to source reduction types: CH, chemical substitution; EQ, equipment change; OP, operational change; PR, product change; PS, process change.
A blank indicates that the plant did not provide information.

Source Reduction Activities

The three source reduction activities that Colloids reported to INFORM for this report are described below and a summary of these can be found in Table II-9. This facility reported no source reduction activities for INFORM's 1985 *Cutting Chemical Wastes*.

Since no hazardous wastes are generated at this plant, the source reduction practices involve nonhazardous waste. In one, a dust collector was installed along with a new dry blending tank. At the end of a run, the collected material is emptied and put into a drum and sold as product.

To avoid waste, samples kept of each product batch for 1 year (in case of customer complaints) are returned to similar batches, not thrown away at the end of the year. The fact that the shelf life of anti-foamers is so long makes it possible to return the samples to a batch of similar product. Also, samples are taken during the batch runs to ensure quality control and reduce off-specification batches. These are tested in the lab and returned to the product batch.

To reduce product container costs and waste, one customer of Colloids requires its products to be supplied in reusable Tote bins. These are metal drums holding 2,000 pounds and are returned empty to be refilled. Their use solves the customer's problem of disposing of metal drums. Colloids' other customers do not buy enough of any one product to take advantage of the Tote bins.

Other Waste Management Practices

Two waste management practices were reported which did fall into the category of source reduction.

Washwater from Colloids' tanks and floor goes to the local sewer. The sewage treatment authority has tested it and found no hazardous substances, though the wastewater may contain trace amounts of raw materials and finished product. The flow averages 17,000 gallons per month.

Solid waste from the plant consists of nonhazardous trash such as paper, cardboard, and occasionally metal from metal drums. Approximately 100 cubic yards per year of such material, at a cost of $60 per month, is hauled to a sanitary landfill.

Change in Yield	Dollars Saved	Dollars Spent	Motivation	Comments	Time Needed for Implementation
			Best control equipment when new process equipment was purchased.		
			Customer requires that product be supplied in reusable Tote bins. Other customers do not buy enough of any one product to take advantage of the Tote bins.		

Technical Assistance

The plant manager at Colloids attends the meetings of the employment advisory group of Contra Costa County, a group of municipal and state employees that provides businesses with information on employment and business activities. They meet about once every 2 months and sponsor lectures by experts in different areas, such as labor laws, immigration laws, and new environmental regulations. The manager reported finding the meetings very helpful and receives the minutes if he is absent.

Technical information and assistance is also available from the environmental, safety and health affairs department in Colloids' corporate headquarters office. There is a system for reporting on problems at the plant. The plant manager reported that he is familiar with the technical and regulatory issues surrounding the types of materials handled at his plant because of his background in pharmaceuticals manufacturing and has not had to call on headquarters for much information.

Company Comments on State and Federal Regulations

Generally, this plant reported meeting the regulations applying to wastewater and solid waste disposal and worker safety. The fee applied by the county under its "Emergency Response" program is the lowest category. The fee is based on employment and amount of materials handled and, therefore, could not be lowered through source reduction.

DEF-TEC CORPORATION
(formerly Smith and Wesson Chemical Company, Inc.)
Rock Creek, Ohio

Summary

This medium-sized plant is in Rock Creek, located amid the farmlands of the northeasternmost county in Ohio, and was part of the large Smith and Wesson company for 20 years. Since 1988, it has been part of the much smaller Def-Tec Corporation. This Def-Tec plant synthesizes a form of tear gas known as CS, but the bulk of its operations concern the assembly of riot control equipment related to tear gas use. One reported consequence of being part of a smaller corporation has been a renewed attention to reducing costs at this plant. Any waste is seen as an added cost, and several projects undertaken to reduce costs have resulted in reduced waste as well.

Smith and Wesson, the plant's former owner, did not grant an interview for INFORM's 1985 study. However, Def-Tec did grant an on-site interview to INFORM for this report and conducted a tour of its facility. The plant reported a total of two source reduction activities reducing 2,535 pounds of waste each year.

Products and Operations

Def-Tec's forerunner, Smith and Wesson, established this chemical weapons plant in 1968 and sold the plant to Lear Siegler, Inc. in 1984. In 1986, this plant and three other former Smith and Wesson plants became the Lake Erie Components Company under independent ownership by a group of private investors. In 1988, this plant was bought by the Def-Tec Corporation. Def-Tec synthesizes a tear gas known as CS (another form of tear gas, CN, is purchased from Eastman Kodak); however, the bulk of the Rock Creek operations is light machine work and assembly of riot control equipment related to tear gas use. There are 65 employees at the plant, a reduction of 35 percent since 1984.

Environmental Policy

While the company does not have a formal environmental policy, the plant manager's policy is "not to make waste because it costs money" — a policy based on the plant manager's 11 years of experience at this facility when it was owned by the Smith and Wesson Chemical Company.

Table II-10 Def-Tec Corporation (Rock Creek, OH): Source Reduction Activities

Waste Medium (SR Type) Year	Source Reduction Activity	Specific Waste Reduced (Hazardous or Nonhazardous)	Percent Waste Reduced	Amount Waste Reduced
Water/solid (PS)	Def-Tec changed its product labeling operations from paint on paper to direct ink application (through silk-screening).	Paint waste	88%	253.5 gal/yr (2,535 lb/yr)
Solid (EQ)	A consolidated press was replaced by a pellet-producing machine.			

Key to source reduction types: CH, chemical substitution; EQ, equipment change; OP, operational change; PR, product change; PS, process change.
A blank indicates that the plant did not provide information.

Materials Data Collection

Regular "scrap reports," begun when the plant was owned by Smith and Wesson, list waste generated per unit of product produced. The reports were set up to identify cost reduction measures, which are often source reduction measures as well.

Other Source Reduction Program Features

While the plant manager is the same under the new ownership, he has found that the significant change with regard to source reduction has been a new emphasis on reducing costs. Any waste is of prime concern since this plant has become part of a much smaller company and such costs loom larger. Less attention was paid and delays in approving capital expenditures occurred within the bureaucracy of the larger company. The manager reported that the new owners are not as reluctant to spend money that will result in reduced costs (and reduced wastes).

Source Reduction Activities

While reporting no source reduction activities for INFORM's 1985 study, Def-Tec reported two for this study. They are summarized in Table II-10 and described below.

With a $10,000 investment in a silk screening machine, Def-Tec changed its product labeling operations from one using paint and paper labels to one directly applying ink to the containers through a silk screening process. The change resulted in a reduction of labeling waste of approximately 88 percent. There are 6.5 gallons of ink waste per year with the new silk screening process, versus 260 gallons of paint waste per year with the old process. This process saves labor and material costs by eliminating the purchase of paint, eliminating equipment cleaning operations, and reducing administrative time in buying labels. Also, the new type of labels is preferred by Def-Tec's customers.

Another source reduction measure required a $17,000 investment in a pellet producing machine that replaced a consolidated press. The new machine generates less scrap and less dust and, therefore, produces more product.

Def-Tec estimated that a total of $5,000 per year has been saved through its source reduction practices.

Change in Yield	Dollars Saved	Dollars Spent	Motivation	Comments	Time Needed for Implementation
		$10,000 for silk screening equipment	Saves operating costs and customers prefer the new label.	Savings from reduced cleaning operations, purchases of paint, and administrative time used for keeping track of labels.	
		$17,000	Product yield increases.	The pellet producer generates less scrap and dust.	

Other Waste Management Practices

Besides the source reduction practices, Def-Tec reported on one waste management measure: the establishment of numerous on-site collection sites for the separation of waste at the plant. Waste such as solvents, paints, and scrap material are kept separate in order to reduce the volume of waste classified as hazardous and, thus, to reduce the costs of analysis for manifesting and disposing of the hazardous wastes. Def-Tec qualifies as a RCRA Small Quantity Generator and uses CECOS (a commercial waste disposal company) for disposal of its hazardous waste.

Technical Assistance

The four main sources of technical information available to Def-Tec are: (1) Def-Tec staff; (2) other companies in this product line; (3) CECOS, the plant's commercial hazardous waste disposal company; and (4) the US military.

Technical cooperation exists between companies operating in this field, and trade associations provide cost reduction information on occasion. The Def-Tec plant also had 18 years of access to the resources of a large company, Smith and Wesson. Now that it is a privately owned firm, Def-Tec's plant manager reported, it is free from the layers of bureaucracy it was once under as part of a large corporation and it is easier to undertake changes. Def-Tec's plant manager also believes that small companies that have branched off from established larger firms have the advantage of experience over new small companies just starting up.

Def-Tec also receives technical assistance from the military. The military carries out its own research as well as awarding research contracts to private companies such as Def-Tec.

Company Comments on State and Federal Regulations

Def-Tec reports that its relations with the state are good and that the Ohio EPA is helpful. While the manager welcomes state source reduction programs, he identified the primary motivating factors for source reduction as profits and market forces.

DOW CHEMICAL USA
Pittsburg, California

Summary

Dow Chemical's Pittsburg, California plant, built in 1916 and located in a highly industrialized area 35 miles northeast of San Francisco, is one of the largest chemical plants in the western United States. It employs 720 people and manufactures more than $225 million worth of products annually. The facility is actually a large complex of several interrelated chemical plants manufacturing both organic and inorganic chemicals. Some operations produce chemicals used as raw materials in the other manufacturing operations at the site. The Chlor-Alkali Plant produces three such chemicals: sodium hydroxide (also known as caustic or lye), chlorine, and a secondary product, hydrogen. The Per-Tet Plant manufactures dry cleaning and industrial degreasers using the chlorine, as well as methane from nearby natural gas wells. The Latex Plant produces latexes for use in water-based paints, coated papers, textiles, and carpets. The other operations at this site produce fumigants and bactericides.

In 1986, the Dow Chemical Company changed its corporate waste management program from one focused on compliance and control to one emphasizing reduction, recycling, and treatment of all chemical waste in all media. The program is called WRAP: "Waste Reduction Always Pays." The WRAP program has a database used to track wastewaters and solid waste and measure progress in reducing them. These data have been collected since 1984. Tracking of air emissions is not as fully developed, primarily because of the difficulty in setting up a monitoring system. An integral part of the WRAP program is its recognition and reward system for employees who suggest and help implement source reduction projects. Other source reduction program features are also noteworthy, such as a full cost accounting policy and strong corporate leadership.

Dow's Pittsburg plant reduced its generation of hazardous wastewaters and solid waste by 93 percent from 1984 to 1988, while production increased by 30 percent during the same time period. The major impetus for this was both the Resource Conservation and Recovery Act (RCRA) and the state of California requirement that Dow close its on-site evaporation ponds. Dow has found that its most effective techniques for achieving source reduction include improved raw material purity and sampling, on-stream analysis of chemicals, preventive maintenance, and improved catalysts.

Dow granted an on-site interview to INFORM for this study, and conducted a tour of its Pittsburg, California facility. It reported three source reduction activities that reduced 12,160,000 pounds of waste and saved $2,726,000 each year. At the time of INFORM's 1985 study, the plant reported two other source reduction activities.

Products and Operations

The Dow Chemical Company operates 32 manufacturing plants in the United States, including four other major manufacturing facilities, in addition to the Pittsburg plant; its United States sales in 1988 were over $16 billion. Dow also operates 150 plants in 31 foreign countries. The Pittsburg, California plant annually manufactures over $225 million worth of products, more than double the value of products five years ago: employment has increased about 10 percent in the same time period and is now about 720.

The Pittsburg facility is a complex of several interrelated chemical plants that manufacture both organic and inorganic chemicals. The Chlor-Alkali Plant at the site manufactures inorganic chemicals, annually producing about 260 million pounds each of sodium hydroxide and chlorine and 7 million pounds of a secondary product, hydrogen, which it sells or burns on-site as supplemental fuel. These three chemicals are sold as product or used as raw materials in the five other manufacturing operations at the complex.

The Per-Tet Plant at the Pittsburg site manufactures two main products: 22 to 38 million pounds per year of perchloroethylene and 27 million pounds per year of carbon tetrachloride, and the by-product hydrogen chloride. All three are manufactured by a process in which an organic raw material (chiefly methane, from natural gas wells located within 60 miles of the plant) reacts with chlorine from the Chlor-Alkali Plant. Perchloroethylene is a common solvent widely used for dry cleaning and industrial degreasing. Carbon tetrachloride, sold to a nearby Du Pont plant, is used as a raw material in the manufacture of Freon, a chlorofluorocarbon.

The Sym-Tet Plant (an abbreviation for symmetrical tetrachloropyridine) annually manufactures 18 million pounds of chlorinated pyridines, one of which is N-Serve nitrogen stabilizer, which kills bacteria in soil. The Dowicil Plant produces 2.8 million pounds per year of Dowicil, a bactericide. The Latex Plant produces 24 million pounds per year of latexes that have a wide variety of applications in the manufacture of water-based paints, coated papers, textiles, and carpets. The Vikane Plant manufactures 2 million pounds per year of Vikane, an inorganic sulfuryl fluoride compound used as a space fumigant.

The Pittsburg site also houses a large research and pilot plant operation which refines chemical production methods and develops new products for test marketing. In addition to carrying out chemical manufacturing, the complex stores and ships dichloropropene for sale as a soil fumigant under the name of Telone.

Environmental Policy

Dow has had a long-standing environmental program, including a toxicology laboratory which was opened at company headquarters in Midland, Michigan, in the 1930s. Current Dow policy and guidelines emphasize source reduction rather than just pollution management and control, as they did in the past. In 1986, Dow formalized a waste management program called WRAP —"Waste Reduction Always Pays." "Waste reduction" is defined as (1) any in-plant practice or process that avoids, eliminates, or reduces waste; and (2) treatment, reuse, or recycling of any material that reduces the volume and/or toxicity of waste prior to final disposal. It covers all media and all chemicals. Dow's 1989 annual report states that the company uses the following hierarchy of management options: (1) source reduction; (2) recycling; (3) treatment and destruction; and (4) secure landfilling as a last resort.

Corporate guidelines spell out waste management practices to be considered. The corporate Environmental Quality Department provides part of the source reduction and recycling training for the company's production managers.

Each Dow division (the Pittsburg site is the headquarters for the Western Division) has an environmental manager who reports to the major manager who has production responsibility for inorganic products, utilities, and the power plant, and who reports to the division's general manager. The environmental managers generally have production experience, and several have spent from 2 to 5 years in the Environmental Quality Department at Dow's headquarters in Midland, Michigan. The Environmental Quality Department reviews new regulations, develops compliance programs, and provides information to the manufacturing divisions. The on-site environmental manager has staff responsibility and can influence and strongly recommend changes at the plant, but cannot require them except for compliance purposes. The general manager has the ultimate authority to require modifications in the plants. This authority is usually delegated through the major manager to the plant superintendent who normally works with the environmental manager to determine when changes are necessary.

Materials Data Collection

Dow's WRAP program has a database for tracking progress. Data are available starting with the base year 1984 for solid waste and wastewater. Starting in 1985, Dow instituted tracking of air emissions from each process at the Pittsburg plant. Employees have inspected and monitored 10,000 sources of fugitive emissions, such as pumps, flanges, and valves. Dow reported that air emissions are the greatest challenge at this facility because of the difficulty in determining what is being emitted.

The WRAP database tracks pounds of each chemical waste per pound of product. The tracking system is designed to identify the actual sources of waste generated within the process. Dow reports that it is able to do this because its products are produced in bulk and not in small batches. The database normalizes the data to take into account swings in production. There is an annual formal review of waste and source reduction for each plant at the corporate level.

The database is tracked both on a total site basis and, in most cases, within individual plants through identification of individual wastestreams or specific chemicals. Dow policy clearly gives source reduction priority over recycling. Its waste reduction statistics include on-site recycling. Air emissions are tracked on a plant-wide basis, but will soon be tied to processes.

Dow officials look to cost accounting as a central part of their reviewing procedures, with costs for waste treatment and disposal allocated back to the individual process and product. Each process has a plant superintendent who has responsibility for costs. Dow also finds it important to track pounds of material lost even if the loss of the material does not cost a lot because the amount lost may be environmentally hazardous. Currently, measuring devices are used to provide a materials balance in the plant but, because of the high volumes of materials in use, it is often difficult for plant workers to measure the very small differences between large numbers. While specific costs allocated to individual products include lost materials, off-site disposal costs, and costs of accidents and waste clean-up, other costs such as regulatory compliance, insurance, and public/customer relations dealing with waste issues are allocated to factory overhead.

Other Source Reduction Program Features

In addition to having a written source reduction policy, materials balance/materials accounting, and full cost accounting, Dow's Pittsburg plant has also fully or partially implemented the other four source reduction program features tracked by INFORM: leadership, an environmental program, environmental goals, and employee involvement.

Central to the WRAP program is a recognition and reward system for employees who suggest source reduction ideas that reduce waste and costs. Source reduction and recycling projects are reported widely throughout Dow. The program recognizes "avoided costs" (such as liability or avoided manufacturing costs over 15 years) as cost savings, thereby taking a long-term view of the costs and savings of projects. Dow reported that the competitive nature of a recognition and reward system has encouraged employees to look for source reduction opportunities. Rewards and recognition include articles in company newspapers, plaques, recognition dinners, and trips to headquarters or other locations to meet with senior management.

The waste tracking system and documentation of costs are used to identify the highest priority areas, to set goals and timetables, and to measure progress. Progress reports are prepared once a month and are presented to the United States area management quarterly.

Dow reports that some of the most effective ways to achieve source reduction it has found include improved raw material purity, on-stream analysis, improved sampling, preventive maintenance, and improved catalysts. Among the obstacles to source reduction that need to be overcome, Dow listed a natural reluctance to change, an incomplete perspective of the whole picture, conflicts in existing government regulations, lack of familiarity with current technologies, fear of compromising product quality, and fear of collecting proprietary data that may then be made known.

Setting up the most appropriate measure for tracking wastes is essential in identifying source reduction opportunities, and Dow has developed a waste index that allows it to do this. At Dow, the process generating the majority of the hazardous waste on-site does not lend itself to the application of the traditional waste unit ratio (total pounds of waste divided by total pounds of raw materials) because the processes have integrated recycle streams and several products are generated. Both the recycle mix and product mix vary throughout the operation. In addition, the molecular weights of these streams have a very broad band. Therefore, Dow has adopted a waste index that uses just one of the common raw materials as the primary index. Since that material either becomes a useful product or intermediate or it becomes a waste, the waste index based on this material encompasses the entire operation. With the benefit of this index, a team of chemists and engineers has analyzed the process, developed and implemented changes, and achieved a 40 percent reduction in waste in the first 6 months of this application. If production amounts did not change from year to year, this would represent an annual reduction of 820 tons of waste.

Dow's corporate Environmental Quality Department aims to rotate production managers into this department and to see that in the future they all have environmental, as well as manufacturing, experience. Only in this way, according to company officials, will top management be committed to environmental quality, and only with the commitment of top management will the entire company be sensitive to the need for such quality.

Source Reduction Activities

Dow reported on three specific source reduction activities at its Pittsburg plant since INFORM's 1985 report; these are summarized in Table II-12 and described below. Dow had reported two other source reduction activities that were described in the earlier report.

Data from the WRAP database detail the source reduction results from Dow's Pittsburg plant from 1984 to 1988 for RCRA hazardous wastes, including both wastewater treated in on-site evaporation ponds and waste sent off-site for disposal. As Table II-11 shows, overall generation of these wastes decreased 93 percent from 1984 to 1988, while production increased by 30 percent during the same period.

Table II-11 Dow Chemical USA (Pittsburg, CA): Change in RCRA Hazardous Wastes, 1984-1988

Year	Treatment On-Site* (tons)	Disposal Off-Site (tons)	Total RCRA Hazardous Waste (tons)
1984	64,040	8,927	72,967
1985	60,766	4,910	65,676
1986	34,595	5,720	42,137
1987	17,298	5,795	53,093
1988	0	4,975	4,975
Percentage reduction since 1984	100%	44%	93%

* In evaporation ponds

Major incentives for this Dow plant to find ways to reduce the wastewater going to the evaporation ponds were federal and state regulations and the rising costs of off-site waste disposal. Dow closed the Pittsburg plant's on-site evaporation ponds, having first been required under RCRA to do this by 1996 and then required by a California law to do so in 1988.

A 1987 process change involving an acid gas adsorption system eliminated the need to send brine to evaporation ponds. This process change, which cost Dow $250,000, reduces caustic waste by 500 tons per month (12 million pounds per year) and hydrochloric acid waste by 160,000 pounds per year, for a savings of $2.4 million per year.

Previously, the wastestream of hydrochloric acid gas, formed by the reaction between chlorine and organic compounds, was scrubbed with caustic, forming brine: a portion of this brine was sent to evaporation ponds while the rest was used to produce chlorine gas through electrolysis. Now, the hydrochloric acid is first scrubbed with water and then caustic. This stepwise method salvages a portion of the hydrochloric acid waste stream so that it can be reused as a raw material in other parts of the plant or sold as product. It also avoids the formation of sodium chlorate compounds that precluded the in-process recycling of the spent caustic stream. Further, less caustic is needed to convert the remaining hydrochloric acid to brine, and all the brine is used as raw material to produce chlorine gas which, in turn, is used, as mentioned above, to produce chlorinated organic compounds in other parts of the plant.

In an operations change related to the evaporation pond closing, storage tanks at the

Pittsburg site were diked to hold flood levels expected only once in 100 years or material from the rupture of any one tank. Trenches and dikes underneath the tanks capture spills and any runoff from the process area; these are then recycled into the brine plant and chlorinolysis unit for use as process water.

The third source reduction activity eliminated 95 percent of the whitewater waste (white solids trapped in the plant wastewater) from the latex operations. This was accomplished by making a survey of the largest contributors to this flow. Many were eliminated and a recycling system was installed to collect the remaining whitewater in a tank and recycle it back into the latex production process. This avoided the need to coagulate the suspended latex particles and landfill them. The cost savings of this project are $26,000 per year plus $300,000 in avoided capital investment in a coagulation system.

Other Waste Management Practices

In addition to its source reduction activities, Dow also reported other waste management practices, including dry sweeping and recycling.

Housekeeping techniques such as dry sweeping have been substituted to eliminate washdowns of the process area. The savings from this change are 200 tons of reduced wastewater per month and $1 million per year in disposal costs.

Recycling was used to reduce the rate of hazardous wastes being discharged to the evaporation ponds, about 5,000 tons per month (60,000 tons of wastewater per year) at the beginning of 1986. Dow estimates that it cost $3 million to install facilities to eliminate this waste and that operating costs will increase by $0.5 million. The project to recycle wastewater used chlorinolysis to convert organics to gases so that they, along with the resulting salt water, could be reused. This chlorinolysis project brought a reduction of over 65 percent in aqueous wastes at this site between 1984 and 1987, and has since achieved 100 percent. In 1988, this amounted to 2,000 tons per month of reduced aqueous wastes and saved $12 million that year, assuming $2 per gallon in disposal costs. Dow estimates that the long-term avoided disposal costs could be as high as $2 million per month (for not disposing of 1 million gallons per month).

It will also cost Dow $12 million to close the ponds because there is no technology that can recycle the salts and the sludge in the ponds.

Technical Assistance

The Pittsburg plant adapted the chlorinolysis technique from Dow plants in Texas and Michigan. Conversely, the source reduction efforts at the Pittsburg latex operations are being exported to other Dow plants. Generally, Dow managers report, the highly visible WRAP program and network of division waste reduction program coordinators help with a mental attitude encouraging seeking further opportunities for source reduction. However, they also report that technology is often too site-specific to be transferable even within the corporation.

Company Comments on State and Federal Regulations

Dow reported that, because source reduction cuts across media, it is often confusing and difficult to know which regulatory standards apply to source reduction efforts. For example, at the Pittsburg site, Dow had to wait two extra months before it could stop a wastewater discharge to the ponds and recycle the water because a new air permit was also required, even though it had a permit for the vent on the old discharge stream.

Another problem Dow identified with the governmental regulations of air emissions leads to a lack of incentive for source reduction. For fugitive emissions, an estimate is produced according to how many valves and flanges exist. No amount of maintenance or improved operations can change the estimate of the amount of such emissions because the number of valves stays the same.

Another problem with governmental standards occurs, according to Dow, when an

Table II-12 Dow Chemical USA (Pittsburg, CA): Source Reduction Activities

Waste Medium (SR Type) Year	Source Reduction Activity	Specific Waste Reduced (Hazardous or Nonhazardous)	Percent Waste Reduced	Amount Waste Reduced
Water/air (PS) 1987	A process change to an acid gas adsorption system in which the wastestream is first scrubbed with water and then caustic (rather than just caustic) to avoid formation of compounds that would preclude recycling the spent caustic.	Spent caustic HCl (H)	100% of wastewater	500 tons/mo (1,000,000 lb/mo or 12,000,000 lb/yr) 160,000 lb/yr
Water (OP) 1987	Installed dikes to hold floods expected once in 100 years and material from the rupture of any one storage tank.			
Solid (OP) 1987	Survey of largest contributors to flow of whitewater eliminated many. A recycling system was installed to collect the remaining whitewater in a tank and return it to latex production process.	Latex solids	95%	

Key to source reduction types: CH, chemical substitution; EQ, equipment change; OP, operational change; PR, product change; PS, process change.
A blank indicates that the plant did not provide information.

intermediate is isolated from a wastestream for reuse in the process. Under the Toxic Substances Control Act, this may require a pre-manufacture notification (PMN), the type of notice required when a company proposes to manufacture a new chemical. This is a lengthy and time-consuming process, company officials report, and discourages the isolation of the material. Filing a PMN costs $2,500 and takes more than 100 hours of work to prepare. Further, as a result of filing, the company may be required to provide extensive and costly documentation. Filing may also subject the existing process to a more restrictive and cumbersome regulatory climate. Dow officials note that, in cases where small quantities are involved, or existing processes are already in operation, an innovative change may not justify the risk of more stringent rules: it may be easier, cheaper, and actually safer, from a regulatory standpoint, for the company to handle the material as a waste, rather than isolating it for reuse or recycling.

Future

Under state laws regulating air emissions in "Toxic Hot Spots" (AB2588) each Air Pollution Control District in California had to select a list of high-, medium-, and low-priority facilities. The Dow Pittsburg plant is required to do an assessment of the risks to the environment and people posed by its emissions because it is on the priority list for its Air Pollution Control District.

Change in Yield	Dollars Saved	Dollars Spent	Motivation	Comments	Time Needed for Implementation
	$2,400,000/yr	$250,000	Closing of evaporation ponds.	Impurities are not produced in the acid neutralization step, so the brine can be reused to produce chlorine.	4 mo
			Closing of evaporation ponds.	Total containment structures capture any spills/run-off from the process area for reuse as process water in the brine plant and chlorinolysis unit.	
	$26,000/yr plus $300,000 avoided capital investment in coagulation system		Avoid capital cost of new system and landfill disposal.	This avoided the need to coagulate the suspended latex particles and landfill them.	6 mo

E. I. DU PONT DE NEMOURS AND COMPANY
Chambers Works
Deepwater, New Jersey

Summary

Du Pont's Chambers Works plant, built in 1917 and located in southern New Jersey, across the river from Wilmington, Delaware, is one of the oldest and largest chemical plants in the United States. There are 3,500 employees and 500 contractors working at the site which has 45 buildings housing five to six separate business operating units manufacturing close to 750 different products and shipping over 1 billion pounds of product per year.

Du Pont established an official company policy "minimizing waste to the extent technologically and economically feasible" in 1980. A primary incentive for this policy at that time was that the on-site wastewater treatment plant's capacity had proven too small to treat all the waste generated when the Chambers Works plant changed from producing solvent-based dyes to producing water-based dyes. This policy has developed into a program which includes a corporate Waste Minimization and Internalization Committee, overseeing the corporate database, policies, and progress reports, as well as a "waste minimization coordinator" at the Chambers Works site, who is responsible for developing, coordinating, and implementing the site's "waste minimization plans." At Du Pont, "waste minimization" includes source reduction and recycling; waste treatment is a separate (and significant) operation at the Chambers Works plant.

Du Pont has a corporate-wide database on the generation of wastewaters and solid waste dating from 1982. The database includes process wastewaters and recycled materials in addition to RCRA-regulated waste, but does not include air emissions. However, its tracking system does not allow monitoring of waste at the process level.

The Chambers Works source reduction program is overseen by a team of plant personnel from plant management, research and development, engineering management, and Environmental Treatment Services. This team reviews major wastestreams at the plant, performance of the different operating areas, and employee awareness of source reduction opportunities. Du Pont has set goals for every plant to meet; the Chambers Works plant reports progress in meeting the goals as the amount of waste reduced per pound of product. The current goals are to reduce hazardous solid and liquid waste by 35 percent from 1990 to 2000 and reduce toxic air emissions 50 percent from 1987 to 1993.

The company did not grant an interview for INFORM's 1985 study. However, for this report, Du Pont did grant an on-site interview to INFORM and conducted a tour of its Chambers Works facility. The plant reported 13 source reduction activities, mostly involving in-process recycling. They reduce 39,290,000 pounds of waste and save the company $3,755,000 each year.

Products and Operations

Du Pont's Chambers Works plant in Deepwater, New Jersey, one of the oldest and largest chemical plants in the United States, now occupies 619 acres. More than 1,200 products — including Teflon, Freon, neoprene, and Orlon — have been invented or developed at the Chambers Works site. Today, it is a highly technical complex using 3,400 raw materials and over 1,000 intermediates to produce 750 products. There are 45 buildings, 3,500 employees, and about 500 contractors at the plant. Over 75 percent of the 2,000 complex operations at the Chambers Works plant are batch processes, making up about 50 percent of the volume of chemicals produced. Continuous operations produce tetraethyl lead, Freon, and some nitrations.

Du Pont's Chambers Works plant is organized into separate production business units that produce chemicals for textiles, automobiles, agriculture, the building industry, soaps and detergents, and intermediates. Over 1 billion pounds of product are shipped per year from

Chambers Works' Deepwater port, in anything from small boxes to tank cars and ships. Forty-two percent of its business is with other Du Pont plants that use materials produced at Chambers Works in other processes.

The site also has a research laboratory which currently concentrates on research for Du Pont's Chemicals and Pigments Department. In addition to products, it develops waste treatment and control technologies.

The Chambers Works plant is a permitted RCRA treatment, storage, and disposal (TSD) facility. More than 600 customers in Delaware and New Jersey send hazardous waste to this facility, which treats over 85 percent of all hazardous aqueous manifested waste (as regulated under RCRA) handled in New Jersey in its on-site wastewater treatment facility. Facility officials also provide assistance to these other waste generators in reducing their hazardous wastes.

Environmental Policy

Du Pont's policy of minimizing waste to the extent technologically and economically feasible, established in 1980, includes a hierarchy of options: source reduction, recycling and recovery of by-products, and detoxification and destruction (treatment and incineration), with containment (landfilling) as a last resort. In 1990, the policy was rephrased as "we will not make, handle, use, sell, transport or dispose of a product unless we can do so safely and in an environmentally sound manner."

The first objective listed in the Chambers Works' mission statement is to operate its facilities emphasizing continued improvement in safety, health, and environmental performance as the highest priority. Techniques for improving yields and consequently reducing waste have been going on for many years at the Chambers Works plant according to Du Pont. In addition, ways to reduce wastes were also being explored during the 1970s as Du Pont changed from solvent-oriented to water-based dye production and the capacity of the site's wastewater treatment plant proved to be too small.

Corporate environmental policy is developed by the Environmental Quality Committee for the Manufacturing Committee. The Manufacturing Committee has a Waste Minimization and Internalization Subcommittee to help develop more specific policies that can then be implemented by the various corporate departments, including the Chambers Works plant.

Each plant site has a "waste minimization coordinator" who is responsible for developing, coordinating, and implementing a written multimedia "waste minimization plan" for the site. The waste minimization coordinator also reports annually on progress made to the corporate database. Du Pont's "waste minimization program" emphasizes reduction, recycling, reclamation, reuse, and detoxification. It does not include waste treatment, which is a separate but significant operation at the Chambers Works plant.

Corporate policy also includes a "waste internalization program" that requires that all wastes be treated on-site or at other Du Pont sites. This program was begun in 1982 in response to a problem of not being able to rely on commercial waste handlers to properly dispose of Du Pont wastes. Further, if the company used a commercial facility that closed down suddenly, the Du Pont plant might also have to close if it could not handle its own wastes.

Plant managers at Du Pont typically come up through the ranks and are moved around to gain a wide range of manufacturing experience. The plant managers' primary responsibility is safety, health, and environmental protection. This includes quality control as well as public relations. The plant manager also works with a business manager at corporate headquarters in Wilmington, Delaware, whose primary responsibility is sales, profits, and product selection.

The plant manager is also responsible for ensuring that hazard reviews are done for any proposed new processes and may make suggestions, such as a different process or safer raw material. A plant manager can also require source reduction measures even if there are resources to treat a waste on-site.

According to a 1986 *Washington Post* article, Du Pont emphasizes both rising treatment and disposal costs and favorable economics of source reduction as incentives for source reduction. An economic and environmentally acceptable plan for waste management can make Du Pont a low-cost producer and hold the key to the success or failure of many of its businesses.[1]

Materials Data Collection

Du Pont has a corporate-wide (except for Conoco and European operations) database on waste generation which dates from 1982 and currently includes data on solid and liquid waste. Air emissions are quantified through engineering calculations based on the number of valves and flanges, rather then on direct measurement. Wastes are tabulated for the following categories: RCRA wastes, on- and off-site disposal of solid waste, ocean disposal, deep-well injection of wastes, process wastewaters (influent to the on-site wastewater treatment plant), waste fuels (RCRA), and recycled materials. Thus, while most of the wastes tabulated in Du Pont's database are RCRA wastes, it does include non-RCRA categories such as process wastewaters and recycled materials. Du Pont officials state that while air emissions are not part the database, plans are underway to include them.

The corporate database took 20 person-years and $5 million to establish. Its primary purpose is for the generation of reports, both corporate and governmental, that can help officials see where and how waste management can be improved. It can, however, also be used to report on source reduction and to identify potential areas for source reduction as data are established on standard measures for waste generation and on waste generation in different years. While waste generation data are initially collected for each process within each business unit at the Chambers Works plant, periodic measurements are then made at a waste collection tank at each of the business units, rather than at each process. The data are specific to wastestreams or to chemical categories and may refer to mixtures of chemicals. Thus, the plant does not do materials balances at a process or chemical-specific level.

Other Source Reduction Program Features

In addition to having a formal source reduction policy and materials accounting (but not materials balance), Du Pont's Chambers Works plant has five other source reduction program features identified by INFORM: an environmental program, environmental goals, cost accounting, leadership, and employee involvement.

The formal source reduction and recycling program began at the Chambers Works plant in 1983 with the formation of a Waste Minimization and Internalization Committee, a year before RCRA's 1984 requirements for a "waste minimization program." Company officials said it was seen as good business: if source reduction was implemented, then Du Pont would be in a good position when the governmental requirements were established while its competitors might not be. The purpose of the committee was to make all the operating departments aware of costs and opportunities for source reduction.

Today, the Chambers Works plant's source reduction program is overseen by a Chambers Works "environmental leadership team" whose membership includes plant management, research and development management, engineering management, and Environmental Treatment Services business management. A subcommittee, the Waste Minimization Task Force, with members from Environmental Services and Environmental Affairs, helped this committee develop a program for 1987-1988 which included: (1) concentration of efforts on the 75 major wastestreams at the plant, (2) development of a "waste minimization culture" for new and updated processes, (3) performance reviews of the operating areas, and (4) efforts to increase employee awareness of source reduction opportunities. The task force meets biweekly.

For the period 1982 to 1990, the Du Pont corporation focused on its solid and liquid hazardous waste. It set goals to reduce these wastes 35 percent on a wet weight basis, or 20 percent on a dry weight basis per pound of product. The goals refer to the total volume or weight

---

1 *Washington Post*, September 25, 1986.

of the wastestream materials, which include both toxic chemicals and inert constituents such as water and soil. In 1986, a goal of 5 percent per year reduction was added. In 1990, a new goal was set to reduce hazardous (RCRA-regulated) waste by another 35 percent by the year 2000. It has been the plant manager's responsibility to achieve the goals.

Air emissions were not originally included in the above goals. However, in 1990, Du Pont established a new goal of reducing air emissions of chemicals on the Toxics Release Inventory (TRI) list by 50 percent by the year 1993, using 1987 as the base year. In addition, a specific goal was set to reduce carcinogen emissions by 90 percent by the year 2000, with the ultimate goal of totally eliminating these emissions.

The Chambers Works plant reports progress toward the goals as the amount of wastes reduced per pound of product. From 1982 to 1986, the plant reduced the stream going to the wastewater treatment plant and the volume of landfilled solids by 16.9 percent. This overall reduction included a 16.8 percent reduction in aqueous wastes per pound of product, a 56.5 percent reduction in solids from the wastewater treatment plant per pound of product, and other reductions in other parts of the plant. These figures are on a wet weight basis and may include some water conservation efforts.

While Du Pont does provide an allocation of capital funds for source reduction projects each year, waste treatment costs are allocated to each process on the basis of wastes generated. These costs include costs of wastewater, landfilled solids, and air emissions and are based on chemical categories such as total organic carbon (TOC), acids, or solvents per 100 pounds of product. Wastes are estimated based on defined waste "standards"; that is, the amount of waste, such as TOC or solvents, generated for a certain product/process per 100 pounds of product. This "standard" amount of waste is used to assign costs to the product. At first, the standards may be estimates of waste generated but, as the process is improved and measurements refined, the standards are redefined. Waste standards were first used in 1974 or 1975 at the Chambers Works plant.

Allocated costs include capital and operating costs as well as the costs of the environmental staff. Specific costs included are costs of lost raw materials, operating expenses for the on-site treatment plants, transportation costs, costs associated with accidents and waste clean-up, costs for regulatory compliance, and costs for dealing with the public and customers regarding waste issues.

The corporation's cost reduction program gives Quality Achievement Awards, including monetary awards, to about 20 employees a month. These awards have included awards for source reduction projects. However, Du Pont officials say they are establishing an award program specifically for source reduction, since, in the long term, economics is a driving force for source reduction. According to Du Pont, the company's less profitable plants, with high waste treatment and disposal costs, already have the greatest economic justification for source reduction. Corporate goals and award programs can provide incentives for plants operating at higher profit margins where the economic incentives for source reduction might not be as strong.

Monthly meetings of all employees are held at Chambers Works to publicize source reduction ideas and emphasize that every employee can make an impact. A Waste Minimization Committee, composed of environmental coordinators from all business units, acts as an information network and makes recommendations. It is used as a resource by the line organizations and also conducts reviews, tracking progress toward meeting the site's waste minimization plan.

As part of the company's overall training program, the Personal Effectiveness Process (PEP) includes sending employees off-site for two weeks every year for problem solving and communication training. Du Pont officials feel that this helps them take advantage of new technologies and changes, including those in the source reduction area.

Every new plant process must undergo a hazard analysis, which is often a fault tree analysis identifying potential problems and solutions. Other procedures for new process approval, including trial manufacturing requests and test authorizations, are used as reviewing mechanisms for proposed source reduction projects.

Du Pont is now applying the procedures used for safety at the plant to source reduction.

Chambers Works has a central safety/health/environmental committee that has to approve any chemical brought on-site. It operates independently of any business unit. Any operator can shut down an operation if he feels there is a safety hazard. Set rules must be followed to resolve the safety issue before the operation can be started up again. Further, there are employee safety awareness meetings, quality control measures, and evaluation of existing as well as new processes.

The Chambers Works plant has a computerized quality management program, available company-wide, that tracks quality beginning with the raw materials. Insisting that suppliers supply high-quality raw materials, plant officials assert, contributes to source reduction. Manufacturing personnel with line responsibilities within each business unit are responsible for maintaining this database, rather than staff, since they are the ones who can take actions to solve the problems.

The Du Pont corporation is conducting a "waste minimization survey" in all its plants to give plant operators a tool for identifying source reduction opportunities. The survey, which is over 100 pages long, lists the questions and data needed to classify sources of wastes and identify technologies to reduce them. It gives the operators a tool for walking through their facility and identifying problems. It also requests information on source reduction projects already done in order to recognize achievements and INFORM others.

Du Pont's policy on waste internalization also provides an incentive for source reduction. To the extent that the Chambers Works plant can reduce its own wastes, its treatment capacity is available for sale to others. The Chambers Works plant charges other Du Pont plants market prices for treating their wastes.

Source Reduction Activities

Du Pont described 13 source reduction activities to INFORM for this study; these are summarized in Table II-13 and detailed below. The company did not report any source reduction activities to INFORM for the 1985 report, *Cutting Chemical Wastes*.

The first source reduction activity, begun in 1983, and motivated by a desire to improve product quality and reduce waste generation, involves insisting that suppliers provide high-quality raw materials.

Du Pont officials discussed eight in-process recovery/reuse source reduction initiatives at Chambers Works with INFORM. In-process solvent recovery has been a part of process design at the Chambers Works plant for more than 20 years. In more recent years, according to plant officials, a major factor motivating maximum solvent recovery has been the increasing costs of incineration, currently $500 to $1,500 per drum.

At the oxyamines building, in the early 1960s, 10 million pounds per year of 1-butanol were being recovered in a closed-loop process and reused at a savings of $2.75 million. Then, in 1980, 75,000 pounds per year of "still bottoms," a solid waste that used to be landfilled, began to be used as a raw material for producing another product.

At the dimethylaniline building, before 1970, 4 million pounds per year of methanol began to be recovered at a savings of $350,000. Most of the methanol is returned to the same process; the rest is used as a raw material in another part of the plant.

At the speciality intermediates building, in 1985, clean-out solvent recovery began to be used to obtain 750,000 pounds per year of ortho-dichlorobenzene for reuse in the same process, at a savings of $260,000.

In 1985, as part of its program to reduce landfilled wastes and to obtain a cheaper source of chlorine, Du Pont began to redistill still bottom purge streams from its chloroamines process. Twenty-two million pounds per year from this stream are now recycled back into the process at Chambers Works.

Also in 1986, in order to reduce operating costs, an in-process iodine recovery unit costing $1 million was installed at the Chambers Works plant. Eighty to eighty-five percent of the iodine, or 200,000 pounds per year, are now sold as product. Du Pont expects to recoup the capital costs of this project within 3 to 4 years.

Another in-process recovery/reuse effort was initiated in 1985 because New Jersey banned landfilling of material with para-chloroaniline. Du Pont began to reintroduce any off-quality para-chloroaniline flakes back into the production process, using distillation. Du Pont found that minor capital expenses, mainly equipment to unload drums, were needed; 65,000 pounds per year were recycled back to the process. In 1990, this process was no longer in use at Chambers Works.

Since 1988, 250,000 pounds per year of waste methyl ethyl ketone have been recycled back into the process with overall savings of $120,000 a year. The process improvements were taken to reduce the costs of raw materials and avoid the increasing costs of incinerating the waste. Other process improvements at the hydrogen reduction building cut by 50 percent the use of methyl ethyl ketone (MEK) as a reactant and solvent and reduced MEK wastes by 200,000 pounds per year.

In 1986, the Chambers Works plant reduced a source of nonhazardous solid waste that was landfilled. The plant replaced the filter aid at the wastewater treatment plant with perlite, which improved the filter operation, reducing the waste generated by 50 percent or 1 million pounds a year. While the cost per pound of perlite is higher than the former filter aid, the improved performance has reduced operating costs of the treatment plant as a whole.

Process control improvements to reduce operating costs in the Monastral manufacturing unit, implemented from 1979 to 1982, resulted in savings in the use and disposal of nitrochlorobenzene. The improvements have reduced waste purged from the process by 750,000 pounds per year, or 65 percent.

In 1985, Du Pont began a vent abatement program to reduce process emissions, prevent accidental releases, and lower plant operating costs. The Freon plant operator did some mass balance calculations for the entire plant's Freon production process and discovered significant losses through evaporation of this gas from the tank trucks in which Freon was delivered to customers. By installing compressors, the company reduced vapor emissions while saving its customers vapor lost from the trucks.

A 1982 assessment of the plant site revealed that the wastewater collection system, then an in-ground open ditch, was a possible source of soil and groundwater contamination and of evaporative losses. All ditches were scheduled to be closed by 1991 and a new collection system, consisting of an above-ground closed pipe, was planned to replace the open ditches and conform to environmental regulations. The cost of closing the ditches was over $10 million. With the new collection system in place, Du Pont expects further source reduction projects to be identified because, for the first time, it will be possible to measure and confirm wastewater generation from specific plant areas.

Other Waste Management Practices

In addition to its source reduction activities, Du Pont reported employing a variety of other waste management procedures. For example, in the production of Freon, which is used in the silicone industry, Du Pont recovers both Freon and hydrochloric acid from the wastestream for reuse. Some 70 million pounds (80 to 85 percent) of hydrochloric acid are now recovered per year. Half is sold and the remainder is used in the ethyl chloride manufacturing process at the plant. The dollar savings are estimated at $6.5 million a year. In 1985, Du Pont spent about $2 million to improve the quality of the hydrochloric acid by-product, enabling the company to provide a higher quality product to its customers. While this did not result in increased reuse or sale of hydrochloric acid, the higher quality product was a key to staying competitive in the production of Freon.

Du Pont built a wastewater treatment plant at the Chambers Works site in 1974. At the time, Du Pont did not concentrate on ways to reduce the wastewater flow but developed a new technology for treating industrial wastes called PACT (powdered activated carbon treatment). PACT removes more than 90 percent of the EPA's organic "priority pollutants" for wastewater. In 1979, when the dye business at this plant was closed down, Du Pont began to market its wastewater treatment capacity (40 million gallons per day) to others. It now handles about 85

Table II-13 Du Pont (Deepwater, NJ): Source Reduction Activities

Waste Medium (SR Type) Year	Source Reduction Activity	Specific Waste Reduced (Hazardous or Nonhazardous)	Percent Waste Reduced	Amount Waste Reduced
Various (OP) 1983	Du Pont insists that suppliers supply high-quality raw materials.	Various (H, N)	Not quantified	
Water (PS) early 1960s	Butanol used as a solvent in oxyamines building is closed-loop recovered and reused.	1-Butanol (H)		10,000,000 lb/yr
Solid (PS) 1980	Still bottoms from oxyamines building are used as a raw material in another product.			75,000 lb/yr
Water (PS) pre-1970	Methanol, a solvent used in the dimethylaniline building, is recovered and reused.	Methanol (H)		4,000,000 lb/yr
Water (PS) 1985	In the speciality intermediates building, cleanout solvent recovery is used to recover ortho-dichlorobenzene.	ortho-Dichlorobenzene (H)		750,000 lb/yr
Solid (PS) 1985	Still bottom purge streams are redistilled from the chloroamines process and recycled.			22,000,000 lb/yr
Water (EQ) 1986	An iodine recovery unit installed in the process. Recovered iodine is sold as product.	Iodine (N)	80-85%	200,000 lb/yr
Solid (PS) 1986	Off-quality parachloroaniline flakes are distilled and reused in the same process.	Parachloroaniline flakes (H)		65,000 lb/yr
Water (PS) 1988	(1) Process improvements reduced use of methyl ethyl ketone (MEK); (2) waste MEK recycled back to process.	MEK (H)	50%	1) 200,000 lb/yr reduced 2) 250,000 lb/yr recycled to process
Solid (CH) 1986	Perlite replaced another filter aid, improving filter performance of wastewater treatment plant.	Filter aid (N)	50%	1,000,000 lb/yr
Water (PS) 1979-1982	Process control improvements in Monastral process reduced use of nitrochlorobenzene.	Nitrochlorobenzene	65%	750,000 lb/yr
Air (EQ) 1985	Compressors installed in tank trucks reduce vapor losses.	Various (H, N)	Not quantified	
Solid/water/air (EQ) 1982-1991	Wastewater collection system (closed, above-ground pipe) being built to replace open ditch system.			

Key to source reduction types: CH, chemical substitution; EQ, equipment change; OP, operational change; PR, product change; PS, process change.
A blank indicates that the plant did not provide information.

Change in Yield	Dollars Saved	Dollars Spent	Motivation	Comments	Time Needed for Implementation
			Higher product quality and lower waste generation.		
	$2,750,000/yr		Cost reduction.		
			Increased costs of landfilling and process yield improvement.		
	$350,000/yr		Operating costs reduction.		
	$260,000/yr		Operating costs reduction.		
			As part of program to reduce landfilled waste banned in New Jersey.		
	$275,000/yr	$1,000,000	Operating cost reduction.	Expect to pay back costs within 3-4 years.	
			New Jersey banned landfilling of this material.	Minor capital expenses, mainly for equipment to unload drums.	
None	$120,000/yr		Operating cost reduction (raw material costs plus reduced cost of incineration).		
			Operating cost savings plus improved filter performance.	Cost per pound of perlite is higher than former filter aid.	3 mo
			Operating cost reduction.		
			Reduced air emissions and operating cost reduction.	Done as part of Du Pont's vent abatement program, result of mass balance calculations.	
		More than $10,000,000 (projected)	Environmental regulations.	Will eliminate evaporative losses and possible soil or groundwater contamination.	

percent of the manifested aqueous hazardous wastes sent to treatment, storage, and disposal (TSD) facilities in New Jersey. Up to 50 percent of the total organic carbon (TOC) load at the treatment plant comes from outside, while 4 to 10 percent of the acid load does. The charge is $1.90 per pound of TOC, $0.34 per pound of total suspended solids (TSS), and $0.33 per pound of acids.

The Chambers Works plant has an on-site landfill that is designated "secure." Its leachate is collected. There is also a decontamination furnace that detoxifies scrap metal for recycling. The plant rinses chemical shipment drums and sends them to a commercial barrel recovery facility.

Chambers Works recovers some chemicals and ships them for use at other Du Pont plants, while it buys chemicals from other company plants. A 2,3-dichloroaniline by-product (1,250,000 pounds per year) is shipped as a co-product to another Du Pont site where it is used as a heat and chlorine source in the manufacture of titanium oxide. The company had to run a year of tests to obtain the government permits to use this material.

Technical Assistance

In recent years, Du Pont has changed its practices with regard to the public. Company officials cite community right-to-know laws as one important reason for this change. Since these laws were enacted, they say, visitors are welcomed and more information is shared with the community and plant workers.

Over 13,000 visitors come to Chambers Works each year. Some are local community residents, but many are others who seek technical assistance on hazardous waste management. The Chambers Works plant's Environmental Products and Services Committee is set up to provide technical expertise on source reduction and waste management to outsiders, in particular to customers in the automobile and plastics industries, but also, for example, to municipal officials seeking information on the storage and handling of hazardous wastes. Also, because this plant is the single available place for disposing of aqueous hazardous wastes for many facilities and communities in New Jersey and Delaware, Du Pont gives assistance and advice on source reduction to these customers.

Du Pont's Chambers Works plant has a Transportation and Emergency Response Team with equipment to handle chemical transportation emergencies. The equipment includes a truck with an on-board computer that can access any Materials Safety Data Sheet as well as the SAFER computer system that predicts the spread of a chemical in the case of an accident. This truck is available to go to an emergency site in the case of a chemical spill.

In 1986, the Du Pont corporation sponsored a symposium on waste management for its employees. Twenty-five outside vendors, including many advocating source reduction techniques, presented their technologies. A similar symposium was held in November, 1988. Du Pont does not use waste exchanges with companies that can use the waste as a raw material, citing the difficulty in having the type and quality of chemicals needed available on a timely basis.

Company Comments on State and Federal Regulations

In Du Pont's experience, some regulations discourage source reduction. For example, one of its Toledo facilities found that recovering solvents would require the plant to obtain a permit as a treatment, storage, and disposal site under the new RCRA definition, so the project was not instituted.

Du Pont has also found problems in the area of air permitting. New permits are required for each new source, for each replacement or modification of equipment, and for new processes that emit fewer air pollutants. Each individual stack or vent needs a permit. Company officials believe that the time needed for approval of new air permits (6 months to over a year) is often too long to competitively introduce a new process, one that may achieve source reduction. In particular, Du Pont has found the time it takes to obtain air permits to be a limiting factor for its agricultural chemicals, since it must react quickly to the market. Du Pont officials view the state air program's resources as inadequate, especially since new chemicals (hazardous air pollutants) and new sources (gas stations) are being added, requiring more staff to review permit applications, but the funds needed for increased staff have not been provided. Currently, the Chambers Works plant has 73 air permit applications pending.

Future

Du Pont plans for a computerized waste monitoring system attached to eight holding tanks, each of which holds chemical waste from five or six buildings and about 500 different processes a day. Based on Du Pont's experience with this type of metering for energy conservation, the company believes that, even though this monitoring system will operate on an area-wide basis, rather than tracking wastes to individual processes, the expertise of the operators combined with the monitoring data will allow the company to find the waste sources.

EXXON CHEMICAL AMERICAS
Bayway Plant
Linden, New Jersey

Summary

Exxon Chemical Americas, a division of the Exxon Chemical Company (which is wholly owned by Exxon Corporation), is the fourth largest chemical company in the United States. It operates the Bayway chemical plant on the same site as one of Exxon's petroleum refineries in Linden, New Jersey. The plant makes olefins (used as raw materials for plastics and other chemicals), additives for fuels and lubricating oils, and some speciality chemicals, and employs about 500 people. One of the oldest plants in the industry, operating since 1921, it is situated in a heavily industrialized area of northern New Jersey, within 10 miles of New York City.

Exxon has no formal written policy on source reduction. The corporate management, Exxon Chemical Americas, has an environmental affairs office that coordinates Exxon's response to federal regulations. The focus is on treatment and control of waste as required by federal laws, and directives are developed separately for air, water, and solid waste, also reflecting federal rules.

The Bayway plant has its own environmental department responsible for compliance with the state and federal laws that affect this plant. Waste management at Exxon's Bayway plant is also directed separately for each environmental medium. Source reduction was incorporated into the management of solid waste in 1985 and 1986, but the issue of how to best set up an all-encompassing source reduction program at Exxon is still under debate.

Exxon granted an interview for INFORM's 1985 study and at that time reported four source reduction activities reducing 681,810 pounds of waste and saving the company $205,305 each year. Exxon again granted an interview to INFORM for this report, with corporate officials from Exxon Chemical. For this study, the plant reported five additional source reduction activities, reducing 16,408,000 pounds of waste and saving the company $3,207,000 each year, none of which was implemented after 1984.

Products and Operations

Exxon Chemical Americas' Bayway plant is located next to a larger petroleum refining operation at the same site. The chemical plant currently has 500 employees, a decrease of about 10 percent since 1984.

The plant manufactures three major product lines:

- Olefins, including propylene and butylenes, used as raw materials in the manufacture of other industrial chemicals and plastics.

- Paramins additives, Exxon's trade name for its line of chemicals that, when added to fuels and lubricating oils, inhibit corrosion, improve flow properties, stabilize viscosity, and otherwise enhance their quality. The Paramins operation manufactures dispersant additives (with the raw material maleic anhydride), detergent inhibitor additives (with the raw material phenol), and synthetic lubricating oils.

- Speciality chemicals, including such isobutylene polymers as LM Vistanex, Exxon's trade name for an ingredient in chewing gum and surgical adhesives.

Since 1988, the plant has discontinued the manufacture of organic solvents.

Environmental Policy

Exxon Chemical Americas has no written policy favoring source reduction. Its Solid Waste Management Plan addresses reduced generation of wastes, recycling, treatment, and safe disposal, but does not put source reduction as its highest priority. The company's corporate environmental affairs office in Houston, Texas, sets broad corporate policy and goals and renders assistance to the individual Exxon plants. It also coordinates Exxon's response to federal environmental laws and regulations. Its written directives include air, water, and solid waste management.

The Exxon headquarters office developed the overall corporate Waste Management Plan in 1982 and 1983. With prime attention to solid waste, it has three goals: to decrease land disposal, to increase recycling, and to ensure that the disposal contractors it uses are competent and responsible. The Bayway plant does not have on-site land disposal or incineration and is not a treatment, storage, and disposal (TSD) facility, so the reduction of solid waste and reliability of its contractors are major considerations.

In general, Exxon's corporate structure is decentralized. The Bayway plant has its own environmental department, which is responsible for compliance at the plant. Because each state's laws differ, individual plants must respond to changes in state laws. New Jersey state laws are the controlling regulations for Bayway.

Waste management at Bayway is directed separately for the separate media. Source reduction was overlaid onto the corporation's management of solid wastes in 1985 and 1986. The use of Waste Minimization Committees, primarily concerned with RCRA compliance, is growing at Exxon plants. The Bayway chemical plant and its adjacent refinery used a Waste Minimization Committee to analyze and follow up on joint problems and opportunities during 1986 and 1987. Currently, the plant has an active Waste Minimization Committee and a management level person in charge of solid waste reduction at the plant.

Nevertheless, Exxon does not currently have a company-wide source reduction program. Officials say they are studying what the best structure may be. A multimedia approach is definitely to be included but the company has not determined just how communication with employees would be done, what type of database would be used, what kind of program top management would be comfortable with, and exact definitions and terminology.

Materials Data Collection

With solid waste and RCRA compliance as its main corporate concerns, Exxon has conducted an annual survey of solid waste since 1982. The survey is carried out at Bayway, as at all company facilities, and is done for each wastestream. The survey gathers information about waste generation rates, disposal locations, and costs for each wastestream, and reports units of waste per unit of product.

The Exxon corporate headquarters uses its annual plant survey of waste to highlight the 20 largest wastestreams (in terms of costs and volume). The corporate environmental affairs department then focuses its efforts on these wastes. None of Bayway's wastestreams were among the largest in the 1987 survey.

The Exxon Bayway plant initiated its own additional plant-wide "waste minimization and compliance review" in 1987, using internal reviewers. It inventoried all wastestreams (solid, water, and air) and reviewed government regulations, especially New Jersey's, to monitor plant compliance.

Cost accounting at the Bayway plant includes solid and aqueous wastes and is directly tied back to the production unit. Air emissions are considered direct raw material or product losses. Specific costs allocated include material losses, disposal costs, costs of regulatory compliance, insurance, clean-ups, and costs of public/customer relations dealing with waste issues. The cost accounting system encourages plant operators to focus on costs and to identify ways to reduce the costs of waste management, including source reduction measures.

According to Exxon Chemical Americas, mass balance accounting is not typically used at the Bayway plant for environmental control because the production volumes are so large that, even though the company's instruments can measure product flow to within 0.25 percent error, that small error could still mean large variations in predicted product flow and, hence, in predicted waste volumes, which are very small in relation to overall product volume. Company officials find it much more effective to directly measure or estimate wastestreams.

In 1984, Exxon Chemical Americas conducted a risk assessment of solid wastes produced at all its plants. As a result, a prime goal of the company's waste management plan is to significantly reduce untreated waste going to landfills. Source reduction, internal recycling, and incineration were reported as leading options for meeting this goal.

Other Source Reduction Program Features

In addition to having materials accounting and cost accounting (but not a formal source reduction policy or materials balance), the Bayway plant has three of the other four source reduction program features identified by INFORM: leadership, an environmental program, and environmental goals, but no employee involvement.

Prior to the 1980s, source reduction at Bayway was considered a process design function where optimization of processes was concerned with material and product losses and the economics of production. Process engineers at Exxon did not generally see environmental operations as part of their role. Exxon's focus was on the "end of the pipe," with environmental specialists whose duties covered treating wastes in an environmentally sound way. In the 1970s, after the price of crude oil rose dramatically, concern increased regarding losses of raw materials. But in the 1980s, source reduction has been encouraged in the solid waste area because of skyrocketing disposal costs (up to a 500 percent increase for some of Exxon's wastes).

Source Reduction Activities

The Exxon Bayway plant reported five source reduction activities (all affecting RCRA waste) for this study, in addition to four other activities the plant had revealed to INFORM for the 1985 study. The five newly reported activities accomplished a total reduction in solid hazardous waste of 18.1 million pounds a year. They entailed an investment of $18.8 million and have produced annual savings of $3.2 million. These activities are summarized in Table II-14 and are discussed in detail below. All of the five newly reported activities were implemented in 1984 or earlier.

The first activity was a project to reduce acid coke, a residue from many processes at the plant, in order to reduce potential long-term liability and disposal costs of the residue. Acid coke is a by-product of the manufacture of butyl alcohol and is a carbonaceous solid saturated with sulfuric acid. Source reduction was achieved through optimization of operating conditions such as temperature and pressure. The study to optimize the process took 5 years, from 1980 to 1984, and was undertaken by the plant engineers. Waste volume was reduced 90 percent from 175 tons in 1980 to 18 tons in 1986, and unit of waste per unit of production dropped from 83 to 7. Prior to 1984, the waste was landfilled; currently, it is incinerated. This measure saved $340,000 per year in incineration charges from 1984 to 1986.

The second source reduction activity cited by Exxon involves a spent catalyst—an organic resin saturated with phenol and a hydrocarbon. The catalyst is used in a batch alkylation reactor until its activity drops below an acceptable level. It is then emptied from the reactor; prior to 1985, it was disposed of in a landfill, but now it is incinerated. Exxon's goal was to reduce the amount of spent catalyst in order to reduce the downtime for catalyst changes as well as to reduce disposal costs and long-term liability. A study undertaken in 1983 to reduce the catalyst waste considered both substitution of other catalysts and process optimization. No replacement catalyst was found, but continuous process optimization of the alkylation reactor operating conditions has extended the useful lifespan of the catalyst by 200 percent and study of ways to further improve the process is continuing. No capital expenditures were required for implementing this project. The waste has been reduced by 11 tons per year, and disposal cost savings are $14,000 per year. Current costs of disposing of this waste are $9,000 for 7 tons of waste. The unit waste per unit of production has been reduced from 19 in 1983 to 6 in 1986, an approximately 70 percent reduction.

The third source reduction activity involved waste oil containing phenols. Prior to 1984, all safety valve releases from a manufacturing unit that contained phenol and an aliphatic hydrocarbon were directed to a blowdown tank where the release was brought into contact with oil in order to capture the vapor/liquid. When the oil became saturated with phenol, it had to be replaced; the waste oil was sent to fuel reclaimers. In 1984, the unit section supervisor suggested that instead of using oil as the contacting medium in the blowdown tank, one of the hydrocarbon raw materials could be used. When the phenol concentration increased above a certain level in the blowdown tank, the hydrocarbon and phenol mixture could be recycled as feedstock to the unit. By substituting a raw material for oil as the absorbing material, the system became closed and the wastestream was completely eliminated. No capital investment for the change was required and the waste elimination has saved $83,000 per year in disposal costs, while reducing 240 tons of waste oil. This project was undertaken to reduce the amount of waste sent off-site for disposal and to save raw material costs.

The fourth example of source reduction at the Exxon Bayway plant is a long-term project (first begun in 1972 and continuing today) to reduce filter cake and process solids generated from the production of lubricating oil additives. After the additive active ingredients are produced in reactors, they contain low levels of solids that must be removed. This had been accomplished by filtration with diatomaceous earth filter aids. The filtration produced a filter cake waste that was 50 percent solids and 50 percent oil and oil additives. The large volumes of solid waste mixed with product indicated to the plant operators that there were opportunities to reduce product losses and disposal costs.

The project began with the replacement of certain filters with high-speed centrifuges that could remove the solids from the lubricating oil additives without the addition of filter aid. The sludge of solids removed by the centrifuges was then blended into waste oils and used as an alternate liquid fuel in industrial furnaces off-site. A series of full-size centrifuges has been installed since 1972. In 1984, second-stage separation devices were installed to recover the oil and active ingredients that remained in the centrifuge sludge. The sludge from the second-stage separation devices was akin to the original filter cake except that it contained no filter aid and was less than half of the equivalent volume of filter cake. Study of possible process modifications, such as extraction, to eliminate or reduce the amount of solids generated in the reactors continues.

According to plant officials, the project used equipment on the leading edge of technology, and assuring consistently high product quality was a concern. The product had to meet high government specifications and there was concern about haze (decreased transparency of the product) after the process changes. Therefore, research on improving product quality and efforts to assure customer acceptance were necessary. As a result, there was a long period of experimentation in which the separation process and mechanical reliability were optimized. The possibilities of recycling the centrifuge sludge to a raw material supplier are also being explored.

The cost of this project has been $18,700,000 since 1972 and has realized a raw materials savings of $1,300,000 per year and disposal cost savings of $261,000 per year. The remaining wastes continue to be landfilled, but the volume of waste has been reduced to one-third the original level (from 5,574 tons in 1980 to 1,478 tons in 1986), for a reduction of over 4,000 tons per year of solid waste. The unit waste per unit of production has been reduced from 105 in 1980 to 34 in 1986, or by 68 percent.

The final source reduction activity cited by Exxon concerns the Bayway plant's production of additives used in lubricating oils. Waste lubricating oil additives are generated when lines and tanks are flushed to avoid cross-contamination when oil leaks from valves and pump seals and when a batch is off-specification. Most of the waste lubricating oil additives are collected from the plant's wastewater collection system. These wastes were reduced both through a study of the lubricating oil manufacturing process and through increased operator awareness resulting from that study. Some of the source reduction measures included better housekeeping and the scheduling of longer campaigns; that is, producing a larger volume of an individual product at one time so that fewer equipment washings are needed. The volume of this waste has been reduced by 50 percent from 1980 to 1986 (from 7,570 tons to 3,774 tons) and has annually saved $214,000 in disposal charges and $1,000,000 in raw material purchases. Previously, the lubricating oil additive waste had been disposed of through third-party waste fuel reclaimers. Since 1985, the remaining waste oil additives have been used as an alternative fuel in cement

Table II-14   Exxon Chemical Americas (Linden, NJ): Source Reduction Activities

Waste Medium (SR Type) Year	Source Reduction Activity	Specific Waste Reduced (Hazardous or Nonhazardous)	Percent Waste Reduced	Amount Waste Reduced
Solid (PS) 1984	Continuous process optimization of operating conditions has resulted in reducing generation of acid coke, a process residue.	Acid coke residue (H)	90%	157 tons/yr (314,000 lb/yr)
Solid (PS) 1983	Process optimization has reduced waste catalyst from a batch alkylation reactor.	Catalyst (H)	70%	11 tons/yr (22,000 lb/yr)
Solid (PS,CH) 1984	A hydrocarbon raw material replaces oil as the phenol-absorbing medium in a manufacturing unit. The hydrocarbon and phenol mixture in the blowdown tank is recycled as feed to the unit.	Waste oil containing phenols (H)	100%	240 tons/yr (480,000 lb/yr)
Solid (PS) 1972	Replaced filters with high-speed centrifuges that remove the solids from lubricating oil additives without the addition of filter aid. Second-stage separation devices installed to recover the oil and active ingredients remaining in the centrifuge sludge.	Filter cake process solids (N)	68%	4,000 tons/yr (8,000,000 lb/yr)
Solid (OP) 1980	Reduction in waste lubricating oil additives accomplished through better housekeeping practices and scheduling of longer campaigns (larger volumes of an individual product are produced at one time, reducing the number of equipment washings needed).	Waste lubricating oil additives (H, N)	50%	3,796 tons/yr (7,592,000 lb/yr)

Key to source reduction types: CH, chemical substitution; EQ, equipment change; OP, operational change; PR, product change; PS, process change.
A blank indicates that the plant did not provide information.

kilns, instead of being sold to recyclers or brokers. Exxon does this to reduce potential long-term liability, even though the cost is higher.

Other Waste Management Practices

Besides the source reduction activities, Exxon also described other waste management practices concerning alternatives to land disposal. Exxon's Bayway plant has moved from land disposal to incineration and other treatment where possible. This decision has increased disposal costs in some cases. For example, the decision to incinerate acid coke increased Bayway's disposal costs by over 400 percent from 1984 to 1986 (from $7,000 to $39,000).

Change in Yield	Dollars Saved	Dollars Spent	Motivation	Comments	Time Needed for Implementation
	$340,000/yr		Undertaken to reduce potential long-term liability and disposal costs of waste previously landfilled. Remaining waste now incinerated.	Waste per unit of production dropped from 83 to 7. Savings are in incineration costs.	5 yr
	$14,000/yr	$0	Undertaken to reduce downtime for catalyst changes and to reduce disposal costs and long-term liability. Current disposal costs are $9,000 for 7 tons.	Waste per unit of production dropped from 19 in 1983 to 6 in 1986. Savings are in incineration costs.	
	$83,000/yr	$0	Undertaken to reduce waste disposed of off-site and for raw material savings.	Savings are in disposal costs.	
	$1,560,000/yr	$18,700,000	The large volumes of solid waste mixed with product made it obvious that there were opportunities to reduce product losses and disposal costs.	Waste per unit of product reduced from 105 in 1980 to 34 in 1986; remaining waste landfilled.	
	$1,210,000/yr		Undertaken to reduce potential long-term liability when disposing of waste through a third party.	Waste lubricating oil additives are generated when lines and tanks are flushed. Savings are from reduced disposal costs and raw material purchases.	

Technical Assistance

Within the corporate structure, Exxon has a Research and Engineering Department that provides regulatory advice and environmental design development as internal consultants to all of the company's plants. It also provides assistance to customers and suppliers when asked. Exxon also has a contractor inspection program that inspects contractors, such as the commercial disposal facilities it uses for hazardous waste disposal, to ensure that they are in compliance with regulations. The company sponsors periodic corporate-wide meetings to exchange information about technical advances, considering this to be the best way to disperse ideas on source reduction throughout its plants. However, Exxon's management believes that most technology in the area of source reduction is plant-specific and that other companies or government agencies probably cannot supply the necessary level of detail or expertise.

The corporation held its first waste reduction conference, which addressed source reduction among other waste management strategies, in September, 1987, at Bayway. Plant managers from throughout the international corporation attended to discuss their problems and solutions.

Company Comments on State and Federal Regulations

Exxon reported that the Toxics Release Inventory (TRI) has been a great additional burden because the company had been tracking wastes by wastestream, as required under RCRA, and not by specific chemicals. Also, materials balance was a driving force for establishing TRI, and Exxon believes that a materials balance approach is not effective when there are small wastestreams from large product flows. If the law were to change to require some sort of materials balance, Exxon's view is that this would be a large waste of resources.

Resources for source reduction and recycling at Exxon were stretched to cover TRI requirements in 1987, but the company hoped to be able to redirect these resources. Because there are already massive incentives for source reduction, Exxon does not believe that TRI will be beneficial in this respect. Also, it does not expect to find any surprises from the data. It sees possible benefits as better federal policy and regulations once a better database is in place.

Exxon officials identified the constant changes in regulations, including definitions of waste, as a serious problem. Each time changes occur, resources that might otherwise be spent studying wastestreams must be diverted to rearrange government reports and continuity of databases used in improving waste controls is lost.

Exxon would like to do more recycling and waste exchanges but believes that RCRA permitting rules are too onerous for most potential recyclers.

One change Exxon has made at Bayway and its other chemical plants is to ensure that technical and communication specialists are on its environmental staff so that, as TRI data become publicly available, Exxon is able to answer questions posed by members of the community where the plant is located.

Future

Exxon reported that the company is in a transition period during which emphasis on treatment technology must give way to a broader, more integrated approach to waste management. Officials believe that this will be difficult because such a large company has many specialists (for example, engineers whose sole duty is to ensure compliance with air permits) who know about treatment technology and their separate media problems but who are now being asked to be health and risk specialists and cross-media experts. Control of chemical wastes does not appear to Exxon to be a problem of technology or numerical incentives for source reduction. Indeed, Exxon reported that requirements such as a given percentage reduction, whether internal or legislated, would more likely reduce the options available, rather than stimulate more source reduction. The company identifies the problem as a management one, involving combining individual expertise and experience in process operations and waste handling in separate media with an expanded view extending beyond treatment and control to the full spectrum of management options.

FIBREC, INC.
San Francisco, California

Fibrec, Inc. did not cooperate in INFORM's 1985 study. All attempts to locate this company for this update suggest that it is no longer in business.

FISHER SCIENTIFIC COMPANY
Fair Lawn, New Jersey

Summary

The Fisher Scientific plant in Fair Lawn, New Jersey, built in 1955, employs about 130 people to make over 1,400 reagent chemicals which are sold, primarily in small quantities, to clinical and industrial laboratories. Located in northeastern New Jersey in the Newark/New York City metropolitan area, this company was a family-run operation until it was bought by the Allied Corporation in 1981 and then by the Henley Group in 1986.

The management changes brought about by Allied included the introduction of a computer-based materials handling system which tracks raw materials, products, and product yield and can provide environmental and safety information for each of the company's products. While the primary purpose of the system was to improve yields, it also ties waste to processes. Fisher management reported that, prior to the installation of this system, tracking waste from the batch manufacture of over 1,400 products was not possible. With the computer system and its timely reports, plant operators have become aware of management's commitment to reducing waste and of their responsibility for the waste generated by their processes.

Another change has been the establishment of a group, called the Waste Minimization Committee, in response to New Jersey regulations. Fisher found that, to be most effective, this committee had to have representatives from all areas of the business — accounting and marketing as well as production. This committee reviews all processes and proposed changes for yield improvement and identifies source reduction opportunities.

Fisher granted an interview for INFORM's 1985 study, but had no source reduction activities to report at that time. Fisher also granted an on-site interview to INFORM and conducted a tour of its Fair Lawn, New Jersey facility for this report. This time, the plant reported 21 source reduction activities, reducing 629,669 pounds of waste and saving the company $529,000 each year.

Products and Operations

The Fisher catalogue lists over a thousand products, ordered mainly in small quantities by clinical and industrial laboratories. The plant's large distribution operations serve primarily scientific and educational institutions. This Fisher Scientific plant employed about 190 people in 1981 when it was bought by Allied. In 1986, the Henley Group took it over. By the following year, employment had dropped to 130.

Environmental Policy

Fisher Scientific does not have a formal, written environmental or source reduction policy. Three years after Allied bought the plant, the company hired a manager of environmental affairs and a full-time professional safety engineer. These new officials made several organizational and management changes at the plant, based on a management commitment to do process research aimed at improving yields and reducing waste. The plant manager reported that directives at the plant reflect Fisher's philosophy of regulatory compliance and economic efficiency, as well as a commitment to being a good neighbor by providing a clean environment.

Materials Data Collection

A key management change made at the Fisher plant by Allied was its introduction of a computer-based materials handling system. Development and installation of the plant software began in 1983. It took two years of work by five people and required an investment of about $1 million. A product management system was also installed, costing an additional $200,000. Prior to this change, there was no materials or production accounting system. Indeed, in 1981, former Fisher officials, interviewed for INFORM's earlier report, had stated that materials tracking of over 1,400 products was not feasible.

The computer system, which began operating in 1986, tracks purchases in pounds of material, products, and product yield against goals. It also has materials safety data and can provide this environmental and safety information for each of its products. The system measures yield with respect to input and output and is developing historical data on inputs and outputs, an essential tool for Fisher in identifying problems and yield targets.

The system was installed as a management tool for improving yields. However, the component for tying wastestreams to processes was included so that the plant could easily provide the compliance information required by state regulations and could allocate the costs of waste to the processes generating them. Through the use of the system, Fisher officials stated, operators have become aware of the management's commitment to reducing wastes and have assumed responsibility for wastes generated by their processes.

Plant management added another tool for tracking materials: two new tanks with load cells. Load cells are electronic stress-strain monitors that measure volume and, therefore, weight, so that it is evident at any time just how much material is stored in a tank.

Other Source Reduction Program Features

In addition to implementing materials accounting/materials balance and cost accounting (but not a written source reduction policy), Fisher has fully or partially implemented three of the other four source reduction program features tracked by INFORM: leadership, employee involvement, and an environmental program (but not numerical environmental goals).

At the beginning of 1987, Fisher established a group, called the Waste Minimization Committee, to review all processes and proposed changes for yield improvements and source reduction. The committee was specifically set up in response to state regulations, including a New Jersey form requesting a "waste minimization target," and its focus was on the solid and semiliquid waste regulated under the Resource Conservation and Recovery Act (RCRA).

At first, the committee was composed of representatives from engineering, operations, and research. However, there were many false starts because the committee did not include representatives from all areas of the organization. For example, there were no representatives from accounting or sales: an accountant can identify areas where a waste might be most costly, and a sales representative knows what outlets might exist for selling wastes as by-products. Fisher officials said they learned that the only effective approach for their waste minimization committee is to have a full-spectrum multidisciplinary team analyzing the problems.

This committee had a specific objective of finding ways to avoid the need to landfill or incinerate wastes. One of its first steps was to do a lot-by-lot analysis of the plant's products to enable the plant to sell off-specification batches that were not suitable for laboratories (Fisher's usual customers) to industrial users instead.

In mid-1989, Fisher hired a new director of safety and environmental affairs at the Fair Lawn plant. Within a year, the staff was expanded to include a manager of environmental affairs and a manager of safety. This expanded team has been charged with carefully examining process operations to identify ways to increase efficiency. Furthermore, new chemists brought on staff in both process operations and quality control laboratories have brought fresh ideas for process improvements.

In June, 1990, the plant initiated a system to regularly collect source reduction information from operations managers throughout the plant. The data gathered, including descriptions of source reduction techniques used, amounts of waste reduced and, in some cases, dollars saved, have enabled the plant to better track source reduction progress as well as to provide this information to its workforce and the community, as required by Section 313 of the Superfund Amendments and Reauthorization Act and by state laws.

The plant has an operator training program which includes information on ways to minimize solvent losses and reduce cross-contamination of different solvents so that they do not end up as waste.

In general, the new management at this plant reported to INFORM that it is effecting a transition from a family-run company which was satisfied with a certain level of profits with no information on yields or efficiencies to a competitive plant with information available to make decisions on how to improve yields and reduce wastes. Officials said this transition has not been easy because of the lack of historical data on trends, product growth areas, and process yields.

Source Reduction Activities

Fisher reported no source reduction activities for the 1985 *Cutting Chemical Wastes* study. However, between 1985 and 1990, Fisher was able to implement 21 source reduction activities at this plant, reducing 629,669 pounds of waste and saving the company $529,000 each year. In addition to information on these 21 individual activities, Fisher officials provided data on overall product yield increases and associated cost savings for two solvent distillation operations (see Table II-15); these improvements are attributable to a combination of different source reduction activities, including re-engineering of production procedures, equipment modifications, and addition of process chemists to enhance production supervision. Increasing the efficiency of these operation also reduces waste disposal costs since less solvent waste is generated: it currently costs Fisher $300 per drum to dispose of solvents by incineration.

Table II-15   Fisher Scientific (Fair Lawn, NJ): Increases in Overall Product Yields for Two Solvent Distillation Operations

Product	Past Yield/ Batch	Present Yield/ Batch	Dollar Savings/ Batch*
Hexane	75%	92%	$7,280
Methylene chloride	75%	93%	$7,280
Estimate for generic solvent production	70-75%	87-92%	Variable

* Batch size based on 2,000 gallons; cost estimates based on unit price of $13/gallon for finished product.

The information Fisher provided about the 21 individual source reduction activities is summarized in Table II-16. The activities are discussed in more detail below.

A desire to improve product yields motivated Fisher to initiate two source reduction activities, both involving process changes. One initiative involved the in-process recycling of what were formerly impure sections of the batches from distillation processes. During the distillation process, the first (called forecut) and the last (called tailcut) materials produced are not as pure as the product produced when the process is in full operation. Traditionally, the forecuts and tailcuts were sold as fuel or disposed of as waste. The engineering department found ways to reintroduce these materials into the process so that they no longer become waste.

The second source reduction activity motivated in this way involved the production of acetonitrile. Fisher concentrated on improving the yield of this process because acetonitrile is produced in relatively high volumes and because the process used three pieces of equipment. The change, which involved going from a two-step to a one-step distillation process, has reduced both air emissions and wastewater discharges from the production process. The two-step distillation process used two reactor vessels: chemical additives were added to the second vessel to scavenge impurities and, since the vessel was unpressurized, there were fugitive air emissions. The process change enabling distillation to occur in just one step eliminated use of the second vessel and, therefore, the air emissions. Research to develop this process change took the on-site laboratory 12 to 14 months, and stabilizing the process for an efficient production run took another 4 to 6 months. The Du Pont corporation provided technical assistance.

Providing higher quality products motivated another source reduction activity. The electronics industry requires high-grade supplies from Fisher Scientific to produce high-grade crystals. Fisher is now starting with higher-grade raw materials in order to provide the higher-grade products. As a result of this operations change, any chemicals rejected because they contain too many impurities for the primary electronics industry customers are still of high enough quality to be marketed to other customers as products.

Another group of four source reduction activities focused on reducing solvent waste at the Fisher plant. Solvents once lost as waste are now recovered from various phases of the distillation process and segregated for reuse and resale whenever practical. Implemented through changes in material handling procedures, this operations change required no capital expenditures. In 1987, 1,500 gallons of solvents (5 percent) were recovered, for a savings of $2,250.

In an equipment change, modifications to existing equipment at a cost of $75,000 decreased solvent wastes by 20,000 gallons per year (5 percent) and increased yields by 28 percent. Dollar savings, based on disposal costs of $1.50 per gallon, are $30,000 per year. Additional savings come from the sale of some of the solvent at approximately $0.50 per gallon.

Two other operations changes affecting solvents involved operator training and collection of solvents from routine line flushing for sale and reuse. Implementation of operator training has ensured minimal solvent losses during routine operations and decreased the frequency of cross-contamination of segregated solvents. Costs of this effort have been $3,000, but Fisher could not quantify the savings and increased yield. Collecting, segregating, and consolidating solvents from routine line flushing cost $500 and reduced waste by 3,500 gallons (5 percent) in 1987, for a dollar savings of $5,250.

In a related effort to sell materials previously considered waste as products, Fisher began segregating and consolidating expired product and quality assurance retention samples in 1987. That year, at a cost of $500, 1,000 gallons of waste were reduced (5 percent) for a savings of $1,500.

Fisher management told INFORM that there had been a "severe" problem with community/plant relations at some point in the past, but that there is now a positive relationship. Company officials cited a desire to respond to increased environmental awareness and concern on the part of workers and the community and to reduce costs associated with waste generation as the motivation for 13 of its 21 source reduction activities. Seven were operations changes, and six were process changes.

Two of these source reduction activities involved selling materials previously disposed of as waste. In one, additional containers were provided so the manufacturing and reclamation and salvage departments could keep different wastestreams separate, instead of collecting flammable material from line flushes, packaging flushes, and laboratories into a common drum. The purer material collected in this way can be sold as a lower grade of product, reducing the amount having to be disposed of as waste. The manufacturing department has reduced waste from ten drums per week to four, and the reclamation and salvage department has reduced waste from eight drums per week to one. Overall savings have been 19,200 gallons (153,600 pounds) per year.

In the other effort, drums containing forecuts, tailcuts, and boilouts are now sampled and analyzed for purity and potential resale; in the past, these materials from manufacturing were automatically declared hazardous waste and placed in hazardous waste storage areas for disposal. Approximately 60 percent of drums analyzed are saleable, resulting in a reduction in waste generated of about 9,900 gallons (79,200 pounds) per year.

A series of process changes also reduced waste. In one, improvements in quality control during operations have led to a steady decrease in waste generated from rejected product since it was implemented. Annual savings in 1990 amounted to $490,000, having started at $330,000 in 1987 and adding additional savings of $110,000 in 1989 and $50,000 in 1990. In another, by reducing the amount of reactant raw materials in purifying tetrahydrofuran (THF) Fisher was able to increase the yield of THF and decrease the amount of off-specification solvent by approximately 30 gallons (180 pounds) per batch. This amounted to about 2,200 pounds in 1990, the year in which this source reduction activity was implemented.

Another process change reduced the amount of solvent waste generated in the packaging process. Up until December 1989, automatic line packaging was preceded by a flush to empty

Table II-16 Fisher Scientific (Fair Lawn, NJ): Source Reduction Activities

Waste Medium (SR Type) Year	Source Reduction Activity	Specific Waste Reduced (Hazardous or Nonhazardous)	Percent Waste Reduced	Amount Waste Reduced
Solid (PS)	Reintroduce impure forecut (first material distilled) and tailcuts (tail-end materials) into the process.			
Water/air (PS)	Two-step production of acetonitrile changed to a single-step distillation process, eliminating need to rinse a second reactor vessel and resulting air emissions.			
(OP)	Starting with higher-grade raw materials provides electronics industry customers with higher-grade products.			
(OP)	Solvents are recovered from various phases of distillation process and reused in process or sold as product.	Solvents (H)	5%	1,500 gal/yr (9,350 lb/yr)
(EQ)	Equipment modifications.	Solvents (H)	5%	20,000 gal/yr (124,740 lb/yr)
Water (OP)	Implementation of operator training to ensure minimal solvent losses and decrease cross-contamination of segregated solvents.			
Water (OP)	Solvents collected from routine line flushings are segregated for resale or reuse.		5%	3,500 gal/yr (21,830 lb/yr)
Solid (OP) 1987	Expired product and quality control samples are segregated, consolidated, and sold as product.		5%	1,000 gal/yr (6,250 lb/yr)
Solid (OP) 1989-1990	Flammable material from flushes and labs kept separate for sale as lower-grade product.	(H)		19,200 gal/yr (153,600 lb or 13 drums)
Solid (OP) 1989-1990	Forecuts, tailcuts, and boilouts are analyzed for purity and potential resale as product.	(H)	60%	9,900 gal/yr (79,200 lb/yr)
(PS) 1987	Improvements in quality control during operations led to decrease in waste generated from rejected product.			
(PS) 1990	Reducing amount of reactant raw materials in purifying tetrahydrofuran decreases off-specification solvent.	Solvent		30 gal/batch (2,200 lb/yr)

Key to source reduction types: CH, chemical substitution; EQ, equipment change; OP, operational change; PR, product change; PS, process change.
A blank indicates that the plant did not provide information.

Change in Yield	Dollars Saved	Dollars Spent	Motivation	Comments	Time Needed for Implementation
			New management investigation of ways to improve yields.	Forecuts and tailcuts had been sold as fuel or otherwise diposed of.	
			Improving yield of the process (one of highest volume products) and eliminating the need for three pieces of equipment.	Technical assistance was received from Du Pont.	18 mo
			Electronics industry requires high-grade crystals and therefore high-grade Fisher products.	With higher-grade raw materials, any rejected chemicals are of high enough grade to be sold to other customers.	
	$2,250 in 1987	$0			
+28%	$30,000/yr	$75,000			
		$3,000		Dollar savings realized, but difficult to quantify.	
	$5,250 in 1987	$500			
	$1,500/yr	$500			
			Concern for worker safety and community relations; reduced disposal costs.		
			Concern for worker safety and community relations; reduced disposal costs.		
	$490,000/yr		Concern for worker safety and community relations; reduced disposal costs.		
			Concern for worker safety and community relations; reduced disposal costs.		

(continued)

Table II-16 Fisher Scientific (Fair Lawn, NJ): Source Reduction Activities (continued)

Waste Medium (SR Type) Year	Source Reduction Activity	Specific Waste Reduced (Hazardous or Nonhazardous)	Percent Waste Reduced	Amount Waste Reduced
(PS) 1989-1990	Redesign of packaging line and filler hopper and use of lower-grade material have reduced need to flush solvent-filling machine.	Solvent		90,000 lb/yr
Solid (PS) 1990	Improved in-process monitoring of Freon has increased yields and quality of product, reducing low-grade waste material.	Freon (H)		15,000 lb/yr
(PS) 1990	Improved in-process monitoring of quality of product (methanol) reduces number of times column material used to absorb impurities is changed.	Slurry and methanol (H)		6,000 lb/yr
(OP) 1990	Methanol contaminated with dolomite particles left in still at end of batch when possible.	Methanol (H)		4,000 lb/yr
(PS) 1985	Process changed to eliminate use of oleum (fuming sulfuric acid) as a solvent in processing acetonitrile.	Fuming sulfuric acid (H)	100%	
(PS) 1987	Process changed to eliminate use of oleum (fuming sulfuric acid) as a solvent in processing hexane.	Fuming sulfuric acid (H)	100%	
(PS) 1989	Process changed to reduce use of oleum (fuming sulfuric acid) as a solvent in processing iso-octane.	Fuming sulfuric acid (H)		
(PS) 1990	Process changed to reduce use of oleum (fuming sulfuric acid) as a solvent in processing cyclohexane.	Fuming sulfuric acid (H)	67%	
(OP) 1990	Before, for acids and ethers, a single bottle from a vendor's case was retained as a sample and the rest was sent for disposal. Now, either a single bottle is obtained or the partial cases are collected and sent to another plant for their inventory.	Acids/ethers		800 gal/yr acids and 400 gal/yr ethers

Key to source reduction types: CH, chemical substitution; EQ, equipment change; OP, operational change; PR, product change; PS, process change.
A blank indicates that the plant did not provide information.

Change in Yield	Dollars Saved	Dollars Spent	Motivation	Comments	Time Needed for Implementation
			Concern for worker safety and community relations; reduced disposal costs.		
			Concern for worker safety and community relations; reduced disposal costs.		
			Concern for worker safety and community relations; reduced disposal costs.		
			Concern for worker safety and community relations; reduced disposal costs.		
			Concern for worker safety and community relations; reduced disposal costs.		
			Concern for worker safety and community relations; reduced disposal costs.		
			Concern for worker safety and community relations; reduced disposal costs.		
			Concern for worker safety and community relations; reduced disposal costs.		
			Concern for worker safety and community relations; reduced disposal costs.		

out trace amounts of previously packaged solvent from the solvent filling machine. This flush resulted in 48 four-liter bottles (51 gallons total) of waste solvent. By redesigning the packaging lines and filler hopper and using lower-grade material prior to packaging, the flush waste has been cut to an average of less than six bottles per run. Since there are 25 automatic line runs per month, this results in an increase of 90,000 pounds of product per year and a corresponding decrease in waste of 90,000 pounds.

A process change implemented in March 1990 improved the in-process monitoring of Freon and changed the processing procedure. This has resulted in increased yields of approximately 2,500 pounds per batch or 15,000 pounds annually, an amount which previously was sent off-site to a material broker or to a hazardous waste disposal facility.

Two source reduction activities (one a process change and one an operations change) reduced the amount of methanol waste. In the processing of methanol, carbon and dolomite columns are used to absorb impurities from the methanol. In the past, the columns were changed after every batch. Through extra in-process monitoring, the quality of the methanol is measured and column material is changed only when needed, eliminating approximately 20 changes per year and reducing the slurry waste by about 6,000 pounds per year. In addition, the last of the methanol to run through the still is left in the still following distillation (when possible) instead of putting it in drums and disposing of it as hazardous waste. This saves approximately 4,000 pounds of methanol (containing dolomite particles) per year.

Four source reduction activities eliminated or reduced the use of flaming sulfuric acid (oleum) in specific operations, thereby reducing generation of substantial amounts of hazardous waste. In 1984-1985, approximately 65,000 pounds of oleum were used per year in processing acetonitrile, hexane, heptane, iso-octane, cyclohexane, and pentane, with the largest amount used for acetonitrile and hexane. Process improvements eliminated oleum from acetonitrile processing in 1985 and in hexane processing by 1987. In 1988, the use of oleum in iso-octane processing was reduced by differing amounts depending on raw material quality and, by 1990, oleum had been reduced in the cylclohexane processing by 67 percent.

Together, these process improvements have decreased oleum consumption to approximately 9,500 pounds per year, despite increases in solvent volumes processed. Had the old processes still been in use at the current rate of production, 110,000 pounds per year of oleum would be used. In addition to the reduction of oleum waste, the amount of sodium hydroxide used to neutralize it (creating a hazardous wastestream) has decreased from 20,000 pounds to 3,000. Thus, total annual waste reduction is 100,500 pounds per year of oleum and 17,000 pounds per year of sodium hydroxide.

Fisher's final source reduction activity involved sample bottles of acids and ethers retained from material used in production. Previously, for each load of these materials vendors shipped to Fisher, a full case was held with the sample bottle. The remaining bottles in the case were then sent off-site for disposal. In 1990, this practice was changed. Now vendors are requested to either retain the sample themselves or to ship a single sample bottle instead of an entire case to Fisher for retention. If this is not done, the partial cases are collected and shipped to another plant where they are entered into normal inventory. The new procedure is expected to save 800 gallons of acids and 400 gallons of ethers a year.

In addition to the costs of individual source reduction activities, Fisher spent $30,000 in 1987 for overall program design and management: engineering time, program and equipment design, and project supervision.

Overall, Fisher considers its source reduction program to be still in the developmental stages. The company expects considerably more waste chemicals will be routinely recovered once the projects are fully implemented and that this will be most evident for the solvent wastestreams.

Other Waste Management Practices

In addition to these 21 source reduction measures, Fisher reported other waste management practices.

During the late 1980's, Fisher established elementary neutralization of corrosive liquids (acids and bases) in a treatment vessel prior to discharge to the municipal sewage treatment plant. Also, certain acids and bases are reused for the neutralization of industrial wastewater in an in-line pretreatment system, reducing these wastes by 25 percent. The cost of these changes was $18,200. They reduce sewer wastes by 17,275 gallons a year, for a savings of $94,250.

Fisher is installing a system to separate its process and sanitary sewer discharges. This is not required by the Passaic Valley sewage treatment plant to which it discharges but is being done as a safeguard. Monitors for the wastewater are also being installed in the hope that they will provide information on ways to reduce these wastes.

Technical Assistance

Technical assistance has been received from both the Digital Corporation in the installation of their computerized materials tracking system and from the Du Pont Corporation in the development of the one-step distillation process. However, in general, Fisher management believes the company is "plowing new ground" and must rely on some trial and error to accomplish its goals.

Company Comments on State and Federal Regulations

Fisher reports that its relations with the state regulatory agencies have been relatively smooth. In some remediation work at the site, there was no comparable work to draw on, causing some disagreements as to what was needed, but these were successfully negotiated.

One aspect of government relations Fisher considered less than satisfactory was the time involved in getting a response from the state after submittal of information by the company; in particular, its application to manage RCRA hazardous waste on-site (RCRA Part B).

Fisher officials told INFORM that debate in New Jersey over the state's Pollution Prevention Act (passed in 1991) spurred them not only to ask how the plant could best use end-of-pipe control to reduce chemical releases, but also to take a closer look at the materials they were using and why. The discussions in the state "started the challenge" and the company "got some interesting answers."

FRANK ENTERPRISES, INC.
Columbus, Ohio

Frank Enterprises told INFORM that the Columbus, Ohio site is "no longer a manufacturing facility and operates solely as a sales office." According to the company's 1987 Hazardous Waste Report, "an explosion on November 10, 1986 destroyed the facility and it will not be rebuilt." The company has withdrawn its generator notification and RCRA Part A permit application for hazardous waste management.

HART CHEM/J. E. HALMA
(formerly J. E. Halma Company, Inc.)
Garfield, New Jersey

Summary

In September, 1988, J. E. Halma Company changed its name to Hart Chem/J. E. Halma and moved from Lodi, New Jersey, to the adjacent town of Garfield. The company's product line — solvents, etchants, acids, and cleansers used by semiconductor and transistor manufacturers — did not change. In 1989, it relocated to Pittston, Pennsylvania. Unsuccessful attempts to contact the plant in 1991 suggest that the plant has closed.

The company did not grant an interview to INFORM for either study. There is no indication that any source reduction measures had been taken at this plant.

Hazardous Substances Used

According to the 1986 New Jersey Department of Environmental Protection's Environmental Survey, Part I, the following environmental hazardous substances had been in use at the plant: acetic acid (glacial), acetone, ammonium hydrogen fluoride (solution), butyl acetate, dichloromethane, etching acid liquid (not otherwise specified), formic acid, hydrochloric acid solution, isopropanol, methanol, nickel and compounds, nitric acid, perchloric acid, phosphoric acid, potassium hydroxide solution, sodium hydroxide solution, sulfuric acid (concentration >51%), thiourea, toluene, 1,1,1-trichloroethane, trichloroethylene, and xylene.

Source Reduction Program and Activities

There is no indication from available records that Hart Chem/J. E. Halma adopted any source reduction techniques.

ICI AMERICAS, INC.
(formerly Stauffer Chemical Company)
Richmond, California

Summary

The ICI Americas' plant in Richmond, California, built in the early 1900s, has changed ownership three times since 1985. Most recently, in 1987, it was bought by ICI Americas (part of the British company, Imperial Chemical Industries) which retained only the agricultural chemicals operations at this site. The manufacture of Devrinol, a herbicide, was moved to Alabama, reducing Richmond's operating budget by 40 percent.

ICI Americas' corporate management has a written environmental policy committed to providing each employee with a safe and healthful place to work and not adversely affecting the environment. While corporate policy does not speak directly to source reduction, the Richmond plant, as part of corporate policy, does have a detailed waste tracking system and reports on progress in source reduction. The Richmond plant has a "waste minimization program" that includes a monetary employee incentive program to encourage source reduction and an environmental/quality committee, which includes representatives from production, maintenance, environmental, and engineering departments, to conduct cross-discipline reviews of all plant operations.

Stauffer, the plant's former owner, granted an interview for INFORM's 1985 study and at that time reported four source reduction activities reducing 5,322 pounds of waste and saving the company $266,085 each year. ICI Americas did not grant an on-site interview to INFORM for this report but provided information through a telephone interview. The plant reported three more source reduction activities, reducing 120,244 pounds of waste each year.

Products and Operations

The ICI Americas Richmond plant has been bought and sold three times in recent years. In March, 1985, the Stauffer company, owner of the plant, was bought by Cheeseborough Pond, which sold it to Unilever, a British company, in 1986. Finally, in 1987, the company's 100th anniversary, the plant was purchased by British Imperial Chemical Industries (ICI) Americas, Inc. Following the acquisition, ICI sold the company's speciality chemical operations to Akzo and its bulk chemical operations to Rhône-Poulenc. Rhône-Poulenc retained the right to the Stauffer name. ICI retained only Stauffer's agricultural chemicals operations, including the Richmond plant.

According to the December 9, 1987, issue of *Chemical Week*, ICI, with the purchase of Stauffer, became the fourth largest US agricultural chemical producer and the world's third largest producer of such crop protection chemicals as the weed-killer Vapam and five herbicides (Eptam, Ordram, Ro-neet, Sutan, and Tillam), with sales of $1.5 billion annually.

Under the different ownerships, the overall scope of operations at the Richmond site has been reduced. In 1987, the manufacture of the proprietary herbicide Devrinol was discontinued at Richmond, and the process was moved to Alabama. This operation formerly accounted for approximately 40 percent of Richmond's operating budget.

Environmental Policy

ICI Americas' corporate management has a written environmental policy, but one that does not specifically identify source reduction as the top priority waste management strategy. The written corporate safety, health, and environmental policy statement, as issued by the chairman of ICI Americas states: "It is the policy of our company that every employee is entitled to a safe and healthful place to work and that the company's activities are conducted in a manner that does not adversely affect the public or the environment," and that "all employees will be held accountable for fulfilling these requirements on the job." In addition, policy statements issued by corporate staff provide further guidance regarding waste handling, waste management, and source reduction to all media. Individual operating sites are given the policies regarding safety, health, and environmental affairs. It is the responsibility of site management to develop programs at the plant level that meet corporate standards.

The Richmond plant continuously monitors source reduction progress and informs corporate staff of current projects and future plans on a regular basis.

Materials Data Collection

The plant has a detailed waste tracking system, as do all ICI Americas' facilities. It tracks waste on the basis of specific chemical constituents, which allows for evaluation of potential source reduction opportunities. All costs of operating the Richmond plant, including specific waste-related costs, are distributed back to original producing cost centers. These distributions may be either direct or indirect, but all costs "of doing business" are ultimately allocated back to the originating processes.

Other Source Reduction Program Features

The ICI Americas' Richmond plant has a written waste minimization program that includes a monetary employee incentive program designed to reward effective ideas that are implemented for source reduction. The program, in compliance with corporate policy, addresses waste releases to all media and includes an employee incentive program designed to ensure involvement.

In 1987, an environmental/quality committee was established. This committee includes representatives from production, maintenance, environmental, and engineering departments. The committee makes monthly tours of the facility to address waste handling concerns as well as to identify source reduction opportunities.

This cross-discipline review of plant operations, along with the employee incentive reward program, encourages employees to "put in their two cents worth." Training programs at the plant also emphasize waste handling, source reduction, recycling, handling of hazardous materials (including waste), and minimizing material on site.

At the Richmond plant there are two technical positions responsible for follow-through on the "waste minimization program": the technical superintendent and the plant chemist.

Source Reduction Activities

Stauffer reported four source reduction measures to INFORM for the previous study, *Cutting Chemical Wastes*, and ICI Americas reported three additional activities for this study. In addition, ICI Americas reported overall waste generation down from over 8,000 tons in 1981 to less than 1,000 tons in 1988 despite an 18 percent increase in production. Some of the reduction in hazardous waste has come about because of the elimination of product lines (for example, the use of toluene has been discontinued because of the cessation of production of Devrinol), but ICI Americas has reported several examples of source reduction implemented or being implemented at the Richmond plant.

Table II-17 illustrates the decline in production-related wastes while overall production grew. Nonproduction-related wastes, such as disposal of materials due to recent elimination of numerous buildings and pieces of equipment at the plant, are not included. Table II-18 shows reductions in chemical wastes reported in accordance with Section 313 of the federal Superfund Amendments and Reauthorization Act (SARA). These data represent a different breakdown of wastes than the RCRA waste categories, but some of the off-site transfers could also be included in the RCRA wastes.

Table II-17 ICI Americas (Richmond, CA): RCRA Waste Generation, 1981-1988

Year	Production-Related Wastes			Total Product (tons)	Index*
	RCRA (tons)	Non-RCRA (tons)	Total (tons)		
1981	8,153	281	8,434	17,812	0.47
1982	4,428	255	4,683	17,451	0.27
1983	6,107	487	6,594	14,628	0.45
1984	4,643	421	5,064	19,909	0.25
1985	4,613	446	5,059	17,660	0.29
1986	5,693	1,556	7,249	15,225	0.48
1987	2,285	1,294	3,579	20,365	0.18
1988	45	634	679	21,110	0.03

* Index is the total tons of production-related waste divided by tons of product.
Source: ICI Americas

Table II-18 ICI Americas (Richmond, CA): TRI Toxic Chemical Releases, 1987 and 1988

Chemical	Surface Water Emissions (lb/yr)	Off-Site Discharges (lb/yr)	Transfers (lb/yr)	Total (lb/yr)
1987				
Toluene	650	0	4,796	5,446
Chlorobenzene	454	0	0	454
Sodium hydroxide	0	3	31,000	31,003
1988				
Chlorobenzene	475	0	0	475
Sodium hydroxide	0	3	10,656	10,659

Source: SARA Title III, Section 313, Toxics Release Inventory (TRI) Report Form R for 1987 and 1988.

ICI Americas reported three source reduction activities at the Richmond plant; they are summarized in Table II-19. In its Generator Hazardous Waste Report to the US EPA for 1987, ICI Americas reported that switching to aqueous-based paints from solvent-based paints eliminated 2,640 pounds of solvent waste. Also beginning in 1987, wastewater from chloropropionamide production has been dewatered, yielding trisodium phosphate-dodecahydrate, a material having commercial value to ICI Americas. The amount of waste reduced by this co-product isolation process in 1987 was 95,284 pounds. In the same process, raw materials are now reclaimed from the wastewaters for reuse within the chloropropionamide process, reducing 22,500 pounds of wastes. These two steps together eliminate all wastes from the wastewaters.

Technical Assistance

Manufacturing and formulation concerns at the Richmond plant involve proprietary processes and chemicals. Therefore, ICI Americas considers that it has the greatest expertise relative to specific products. In addition, the Richmond plant shares the same site with the ICI Americas Western Research Center, the sole agricultural research facility for ICI Americas in the United States. This proximity allows for the availability of technical assistance "right up the street."

Table II-19 ICI Americas (Richmond, CA): Source Reduction Activities*

Waste Medium (SR Type) Year	Source Reduction Activity	Specific Waste Reduced (Hazardous or Nonhazardous)	Percent Waste Reduced	Amount Waste Reduced
Solid (CH) 1987	Switching to aqueous-based paints eliminated solvent wastes.	Solvent waste (H)	100%	2,460 lb/yr
Water (PS) 1987	Dewatering wastewater from chloropropionamide production process yields trisodium phosphate-dodecahydrate, a commercially valuable material.	Trisodium phosphate-dodecahydrate (H)	81%	95,284 lb/yr
Water (PS) 1989	Raw materials in wastewaters reclaimed for reuse in production process of chloropropionamide.	(H)	19%	22,500 lb/yr

* ICI Americas did not provide information on yield changes, dollars saved or spent, motivation, or implementation; therefore, the right side included in all the other source reduction activity tables is not included here.

Key to source reduction types: CH, chemical substitution; EQ, equipment change; OP, operational change; PR, product change; PS, process change.
A blank indicates that the plant did not provide information.

Future

An example of waste reduction as a result of the "waste minimization program" and the environmental/quality committee is the system being installed at the plant for rinsing Vapam filter cartridges. The cartridges will be rinsed to reclaim product Vapam and, thereby, will no longer be classified as a RCRA-regulated waste.

ICI RESINS
(formerly Polyvinyl Chemical Industries)
Vallejo, California

Summary

The ICI Resins plant in Vallejo, California, among vineyards near San Pablo, about 10 miles north of San Francisco, was established in 1970 and was originally owned by California Resin and Chemical Company. It was bought by Beatrice in 1982 and by the British company Imperial Chemical Industries (ICI) in 1985. It now employs 15 workers. ICI Resins produces water-based acrylic resins. In 1984, it discontinued the manufacture of alkyd resins, due in part to the state of California's ban on the use of oil-based paints.

ICI Resins' waste management and source reduction program has been developed at the plant site by the plant manager and his technical manager. Under the original ownership, the plant was not very profitable and had few resources for research into environmental controls. Under Beatrice, corporate workshops on hazardous waste management made the plant manager aware of the need for a person on-site to oversee these issues. The technical manager took a course offered by the University of California and has set up a waste tracking system and an inspection system that monitor all hazardous waste. He has the authority to require changes leading to source reduction. While very little change in personnel has taken place at this plant under the changes in ownership, a major effort has been required to train and reeducate plant employees to change their habits and methods of waste handling in order to accomplish source reduction.

Polyvinyl, the division of Beatrice that was the plant's former owner, did not grant an interview for INFORM's 1985 study. ICI Resins did grant an on-site interview to INFORM for this study and conducted a tour of its Vallejo, California facility. The plant reported a total of three source reduction activities.

Products and Operations

The major product at the Vallejo plant is acrylic polymers, which are produced using emulsion polymerization. The resulting lattices are sold to the protective coating, floor care, and cement admixture industries. The basic equipment used at this plant is polymerization reactors, storage, feed and blending tanks, a steam boiler, and a cooling tower. Materials include organic monomers, emulsifiers, free radical initiators, reducing agents, solvents, and special additives. There are quality control and process technology laboratories at the plant.

This plant site was originally owned by California Resin and Chemical Company. Beatrice purchased it in 1982. Polyvinyl Chemicals, Inc. (a division of Beatrice) was bought by Imperial Chemical Industries (ICI) in 1985. The name was changed from Polyvinyl Chemical to ICI Resins in 1988. The current plant manager has worked at this plant under the last two owners.

California Resins and Chemical Company's primary product line was alkyd resins, lacquers, and enamels. After a transition period of less than 2 years, the plant discontinued the manufacture of alkyd resins in June, 1984, due in part to the fact that California banned the use of oil-based paints. ICI Resins' new product line is primarily water-based acrylic resins. With the discontinuance of the alkyd business, materials such as toluene, xylene, maleic anhydride, and methanol were no longer used at the plant. With this change in business, ICI Resins has had to develop a new set of customers. It was not able to persuade its old alkyd customers to change.

Environmental Policy

The ICI Resins plant has a waste management/reduction program designed to provide training, guidance, and support for source reduction efforts. A formal, written policy at ICI Resins' Vallejo plant has been in effect since 1985. The policy lists 35 substances used at the plant that are considered hazardous, and states that all waste (air, water, and solid wastes) generated at this plant must be evaluated to determine if it is hazardous.

The policy emphasizes that sources of hazardous wastes include leaks, spills, rinses, samples, and empty containers. The first item under the hazardous waste control section is source reduction, including preventive maintenance and engineering controls. Hazardous waste control includes recycling back into the manufacturing process. Disposal (solid waste and discharge to the municipal sewage treatment plant) is included under a separate heading.

Materials Data Collection

While ICI Resins reported that materials balance is difficult, it has taken steps in this direction. Inputs can be measured quite accurately, but the measurement of bulk yields on an individual batch basis is less so. ICI Resins measures what it sells by keeping track of the weights of products that go into trucks or into drums. The material can be weighed to an accuracy of 0.5 to 1 percent. Waste generation is calculated as the difference between inputs and product outputs. ICI Resins undertakes quality reviewing to verify yields. The materials balances are done for each product and process.

Other Source Reduction Program Features

Under California Resin ownership, the plant produced alkyd resins used in oil-based paints. California Resin was an undercapitalized plant, not particularly profitable, and the plant manager reported that few resources were available for research into environmental controls. It was Beatrice that first held workshops to discuss hazardous waste. As a result, the plant manager decided he needed a person to oversee these issues so he sent the technical manager to the University of California at Davis to take its hazardous waste management course. The technical manager earned his certificate in 1983 and wrote the plant's policy on hazardous waste handling.

Under this policy, when a waste is generated, a hazardous waste form is filled out by the operator. Only the technical manager can classify the waste as hazardous or nonhazardous and dictate its disposition. The technical manager oversees all hazardous waste issues and has the authority to require changes leading to source reduction of the wastes.

ICI Resins does a weekly inspection for worker safety and environmental hazards, including avoidance of generating hazardous wastes from such sources as leaks and spills and mishandling of products. The plant manager reported that this has helped bring about a change in attitude because the employees do not know when they might be inspected. Scores are posted and the employees compete for the best score.

As part of the hazardous waste management at the plant, goals and timetables are set for the reduction of particular wastes, which can be air emissions, wastewaters, or solid wastes, depending on the process involved.

While very little personnel change has occurred at this plant under three different owners, quite a lot of change in methods of handling wastes has been accomplished, primarily through training and reeducation to change habits developed over the years. The plant manager estimated that up to 75 percent of the source reduction accomplished from 1984 to 1987 is the result of this training. The training is based on the book *Quality Is Free,* by Phil Crosby, which establishes total quality control in all aspects of the operations as a key to a profitable enterprise, a concept designed to make everyone responsible for quality and aware of what they have to do to achieve it.

Source Reduction Activities

ICI Resins reported implementing three source reduction activities for this study. These are summarized in Table II-20 and further detailed below. No source reduction measures were reported in INFORM's 1985 study.

Table II-20 ICI Resins (Vallejo, CA): Source Reduction Activities

Waste Medium (SR Type) Year	Source Reduction Activity	Specific Waste Reduced (Hazardous or Nonhazardous)	Percent Waste Reduced	Amount Waste Reduced
Solid (OP)	The plant has instituted "ABC" inventory control so that stored materials do not go bad before they are used.			
(EQ)	Leaks have been reduced through the use of non-leaking pumps and replaced pipes and hoses.			
(OP) 1985	Materials from leaks that cannot be eliminated are recycled. Catch pans are used underneath trucks during loading. Spills are put into drums along with lab samples and reused in the process.			

Key to source reduction types: CH, chemical substitution; EQ, equipment change; OP, operational change; PR, product change; PS, process change.
A blank indicates that the plant did not provide information.

ICI Resins estimates that 97 percent of its hazardous waste has been eliminated, from 29 tons of hazardous waste shipped off-site in 1983 to an estimate of 1 ton in 1988.

1983	29 tons
1984	130 tons
1985	143 tons
1986	8.4 tons
1987	6 tons (estimate)
1988	1 ton (estimate)

ICI Resins identifies the high amount of hazardous waste in 1984 and 1985 as wastes generated from getting out of the alkyd business. Although about 200 drums of waste were sold, the remainder had to be sent for disposal.

ICI Resins further estimates its source reduction program, begun under Beatrice ownership and continued to the present, has been responsible for eliminating 85 percent of the plant's hazardous waste. He reported that one particularly successful source reduction practice has been improved methods of inventory control. ICI Resins has ABC inventory control which classifies raw materials by quantity used and cost: "A" items are expensive and can be ordered, as needed, on a daily basis; "B" items are used less and cost less and are ordered weekly; and so on. This system was started under Beatrice ownership, primarily to control costs, but it has also reduced raw material waste. The proper handling of raw materials has resulted in fewer raw materials needing disposal.

The other two source reduction activities involve controlling leaks, and accounted for a total of 10 to 15 percent of reduced wastes. A variety of engineering controls include eliminating leaks by changing to nonleaking pumps and replacing pipes and hoses. In addition, material from leaks that cannot be eliminated is recycled. For example, since 1985, catch pans have been used underneath trucks being loaded with product. The spills are put into drums along with lab samples and then recycled back to the process.

Change in Yield	Dollars Saved	Dollars Spent	Motivation	Comments	Time Needed for Implementation
			System started to control costs of raw materials.	Inventory control leads to less spoilage of materials which then have to be disposed of.	
				All leak control measures combined account for 10 to 15% of reduced waste.	

Costs of achieving source reduction have included $20,000 for analytic instrumentation for conducting quality control checks. Operators have been trained to do their own checks early in the process so that bad batches can be avoided. Because just one bad batch can cost over $20,000, the pay-off for this equipment is less than a year.

Other Waste Management Practices

ICI Resins also reported the following waste management measures to INFORM.

ICI Resins' rinsewater is sent to the local sewage treatment plant. The wastewater contains no Clean Water Act priority pollutants except for zinc, which is currently regulated at less than 1 part per million (ppm). In the future, ICI Resins may be required to pretreat the zinc in the wastewater, and it is investigating ways to reduce the zinc so it will not have to be treated. The plant's sewer fee is $1,000 to $1,200 per month, based on flow and chemical oxygen demand (COD). It currently hauls its wastewater to the sewage treatment plant but expects to be hooked into the sewer soon.

The California standard for air emissions from a resins plant is a maximum of 10 pounds per day of volatile organic chemicals (VOC). Less than 8 pounds per day are emitted from this plant. The vapor pressures of materials used at the plant are well known. They are refluxed in a closed system so that materials go back into the process.

In 1990, a system to suppress the release of chemical vapors in case of a spill from the storage tanks was installed. If a spill occurs, this system automatically covers the spill with foam so that the chemical will not evaporate. However, the foam renders the spilled chemical unusable at the plant so that it must be disposed of as waste.

Another new system is planned. A non-RCRA waste will be filtered, thereby creating two streams: a wastewater sent to the public sewage treatment plant, and a low-grade polymer which could be sold as a product for use in paints suitable only for short-term applications. This low-grade polymer could not be reused in processes at this ICI Resins facility.

Technical Assistance

The current owner, ICI, has a department to assist its plants in environmental controls and holds workshops with representatives from each of its plant sites. However, each plant manager remains responsible for the performance of his plant. Because this plant is located in California, which has extensive state and local environmental laws and regulations, many unique to the state, the technical manager at this plant must stay abreast of environmental issues.

Of the trade associations with which the plant manager is acquainted, he reported that a good source of technical information is the Chemical Industry Council of California.

Company Comments on State and Federal Regulations

California's Tanner Act has increased local government's activity in the hazardous waste control area. The law requires each county to describe sources of waste generation and to develop plans for handling this waste within the county up to the year 2000. The plant's technical manager is a representative on the county's Tanner Committee. Napa County is not highly industrialized; ICI Resins is the only chemical company in the county. ICI Resins reported that the Tanner Act has been a good law, bringing in the resources of the local government, which the manager describes as quite sophisticated compared with resources available in most states.

While the local government has been helpful, ICI Resins has found that the state personnel do not seem to have much knowledge of chemical processes. It seems to ICI Resins managers that the state has been swamped by new laws and regulations and appears cautious in its implementation.

Future

The plant's technical manager foresees a problem in disposing of very small amounts of currently exempt waste in the future. Resource recovery works with large-volume wastestreams, but when just a few gallons of waste are left, they may be hard to get rid of. He reports that waste exchanges do not work because it is often too difficult to know what is really in the container.

INTERNATIONAL FLAVORS AND FRAGRANCES, INC.
Fragrance Ingredients Plant
Union Beach, New Jersey

Summary

The International Flavors and Fragrances (IFF) plant in Union Beach, on the New Jersey coast 20 miles south of Newark, is considered a large quantity generator (greater than 1,000 kg per month) of hazardous wastes regulated by the US EPA under the Resource Conservation and Recovery Act (RCRA). This plant manufactures synthetic perfume and flavor chemicals for use in food, cosmetics, soaps, and detergents. Built in 1952, the Union Beach facility produces 400 to 500 products on a routine basis, with a capacity to manufacture about 800 different products. Employment at this facility has decreased from 375 in 1983 to 225 in 1985 ("New Jersey Environmental Survey," 1986).

Under state regulations, the plant reported that it does not have a source reduction policy, but that its 1987 "waste minimization plan" did include source reduction as the preferred waste management strategy. This report to the state also indicated plans underway to establish criteria for selecting processes or wastestreams to study for ways to reduce waste and to establish a database for RCRA waste at the plant. The plant has an employee program of financial rewards for recommendations concerning environmental matters.

International Flavors and Fragrances did not grant an on-site interview to INFORM but provided some information through a telephone interview. In written materials provided to the state of New Jersey, the plant reported a total of 11 source reduction activities averaging about 30 percent reduction in the amount of waste generated for the individual wastestreams affected. The company did not grant an interview for INFORM's 1985 study.

Products and Operations

IFF supplies flavors and fragrances to the food, beverage, cosmetics, soap, and detergent industries worldwide. The Union Beach plant manufactures approximately 800 products, 400 to 500 on a routine basis. Approximately one-third of these products are flavorings; the others are fragrances.

According to the company's 1987 annual report, corporate sales and earnings for that year were the highest in the company's history; increases over the previous year of 20 percent and 25 percent, respectively. The total sales of $461 million in 1983 rose to $745 million in 1987, almost three-quarters of which were outside the United States.

Environmental Policy

According to its New Jersey Hazardous Waste Generator Waste Minimization Report for 1987, IFF does not have a written policy or statement outlining goals, objectives, and methods for source reduction. In 1987, the plant did submit a "waste minimization plan" to the state that included, as the optimal waste management strategy, the hierarchy of source reduction, followed by recycling and material recovery, treatment, incineration and, finally, disposal, as the least desired alternative.

Materials Data Collection

The 1987 "waste minimization plan" states that IFF will establish a database covering solid, liquid, and semisolid wastes, which will measure amount or volume of the waste. IFF reported that developing a manageable database to cover the 400 to 500 products produced on a routine basis at this plant will require discussions with the state. For the year 1988, the waste minimization program expected to establish selection criteria and select production processes and/or wastestreams to be studied based on the criteria.

Because this plant did not grant an on-site interview, INFORM examined government reports to obtain information on waste generation and releases. Amounts of RCRA waste and TRI releases and transfers are presented in the following tables. Table II-21 indicates that the pounds of RCRA hazardous wastes generated and sent off-site for disposal decreased by 7.5 percent from 1984 to 1987. This information by itself does not indicate if this reduction is due to production changes, releases of the wastes shifted to another environmental medium, reclassification of wastestreams, on-site treatment, or actual reductions at source.

Table II-21   IFF (Union Beach, NJ): RCRA Hazardous Waste Generation, 1984-1987

Type	Year	Amount In Pounds	Amount In Gallons
Handled on-site	1984		
	1985	11,477,020	293,242
	1986		
	1987	10,763,472	134,015
Stored at year end	1984		
	1985	581,250	0
	1986		
	1987	647,442	0
Sent off-site	1984	10,942,293	140,364
	1985	11,466,220	89,612
	1986	7,594,998	
	1987	10,116,030	123,933

Note: A blank indicates that the plant did not provide information.

Sources: 1984-1986, New Jersey Department of Environmental Protection, "Hazardous Waste Generator Annual Report"; 1987, New Jersey Department of Environmental Protection, "Treatment, Storage, and Disposal Facility TSDF Annual Report, Form I."

Table II-22 shows the releases and off-site transfers of chemicals reported to the Toxics Release Inventory (TRI) in accordance with Section 313 of the Superfund Amendments and Reauthorization Act (SARA). While the information is based on chemicals used or manufactured at an industrial facility and is broken down differently from RCRA waste, the category of "Off-site Transfers" in Table II-22 could include some RCRA waste.

For TRI, all types of releases, not just solid and liquid wastes handled by treatment, storage, disposal, and recycling facilities (TSDR), must be reported. Thus, for these wastes, it is possible to determine whether wastes have been shifted between environmental media. Also, the state of New Jersey requires reports on source reduction and recycling actions, if any, for these chemicals (see Table II-23, below), so that data are available on whether or not a chemical has been subject to source reduction at the facility.

Table II-22 shows that over 20 TRI chemicals are used or manufactured at this IFF plant in quantities greater than the threshold amounts specified by the regulations (10,000 pounds used or 50,000 pounds manufactured in 1988, and 75,000 pounds in 1987). Off-site transfers were greatly reduced between 1987 and 1988 because of the decrease in toluene wastes, from more than 1 million pounds to about 23,000 pounds. Other large decreases were reported for chromium compounds, ethylbenzene, methyl ethyl ketone, sodium hydroxide, and xylene. The one chemical for which a significant increase was reported is phosphoric acid, which increased by 38 percent, from 150,000 pounds to 207,000 pounds.

These overall decreases in off-site transfers might be decreases in the amount of waste generated at the plant, rather than just shifts to other media, because other types of environmental releases of the same chemical did not increase. However, off-site transfer for recycling or reprocessing does not have to be reported on the TRI form. For the federal TRI reports for 1987 and 1988, the section on source reduction was voluntary and IFF did not indicate that it had undertaken source reduction for any of these chemical wastes. However, data available from the state of New Jersey from 1986 to 1987 indicate that IFF undertook source reduction for 11 of the 21 chemicals (see Table II-24 on page 198).

Table II-22   IFF (Union Beach, NJ): TRI Toxic Chemical Releases, 1987 and 1988

Chemical	Air Emissions (lb/yr)	To Public Sewage (lb/yr)	Off-Site Transfers (lb/yr)	Total (lb/yr)
1987				
Acetaldehyde	250	250	1,250	1,750
Acetone	500	250	2,000	2,750
Acetonitrile	500	0	250	750
Chromium compounds	0	0	82,000	82,000
Cyclohexane	500	250	1,500	2,250
1,2-Dichlorobenzene	500	250	1,500	2,250
Dichloromethane	500	250	1,500	2,250
Diethyl phthalate	500	250	1,500	2,250
Ethylbenzene	500	250	133,500	134,250
Ethylene glycol	500	250	750	1,500
Formaldehyde	500	250	1,500	2,250
Hydrochloric acid	500	0	0	500
Methanol	1,000	250	6,200	7,450
Methyl ethyl ketone	500	250	58,450	59,200
Phosphoric acid	250	0	150,000	150,250
Propylene oxide	1,500	250	1,500	3,250
sec-Butyl alcohol	0	250	37,500	37,750
Sodium hydroxide	500	250	571,250	572,000
Sulfuric acid	500	0	0	500
Toluene	2,650	250	1,008,300	1,011,200
Xylene	500	250	671,950	672,700
Total	12,650	4,000	2,732,400	2,749,050
1988				
Acetaldehyde	500	250	750	1,500
Acetone	500	250	750	1,500
Acetonitrile	500	0	0	500
Butyraldehyde	500	250	750	1,500
Chromium compounds	0	0	500	500
1,2-Dichlorobenzene	500	250	1,250	2,000
Dichloromethane	500	250	750	1,500
Diethyl phthalate	500	250	750	1,500
Ethylbenzene	500	250	1,000	1,750
Ethylene glycol	500	250	750	1,500
Formaldehyde	500	250	750	1,500
Hydrochloric acid	500	0	250	750
Methanol	500	250	750	1,500
Methyl ethyl ketone	500	250	750	1,500
Phosphoric acid	250	0	207,227	207,477
Propylene oxide	500	250	750	1,500
sec-Butyl alcohol	500	250	750	1,500
Sodium hydroxide	500	250	750	1,500
Styrene oxide	500	250	750	1,500
Sulfuric acid	500	0	500	1,000
Toluene	1,354	250	22,103	23,707
Xylene	500	250	1,000	1,750
Total	11,104	4,250	243,580	258,934

Source: SARA Title III, Section 313, Toxics Release Inventory (TRI) Report Form R for 1987 and 1988.

Other Source Reduction Program Features

The 1987 "waste minimization plan" indicated IFF would establish a system for identifying high priority wastestreams to be investigated by groups within IFF for ways to reduce the wastes. The groups would include the participation of research and development staff, process control and production staff, and IFF's department of environmental compliance. The technical investigation portion of the "waste minimization plan" focuses on actions that maximize product yield, evaluate alternative production processes, recycle and recover materials, and reduce volume through on-site treatment.

IFF also sponsors "The Better Way" (or quality circles), which provides a forum for groups of employees to get together to raise and address important issues. This program has resulted in recommendations concerning environmental matters, subsequently implemented by IFF. Both this program and their "Suggestion Award Program" provide for financial recognition to the employees involved.

Source Reduction Activities

Data are available on the overall decrease in certain chemical wastes from 1986 to 1987 at this plant through the New Jersey Right-to-Know program. Table II-23 shows reductions of 3 to 34 percent for ten specific chemical wastes, and an increase of 2 percent for another. These normalized figures are adjusted for increases in production between 1986 and 1987. That is, total waste generation for each of the 11 chemicals cited increased anywhere from 10 to 26 percent, while product output increased anywhere from 20 to 60 percent. For example, acetaldehyde production increased 60 percent, but waste generation increased only 26 percent, resulting in an adjusted reduction of 34 percent.

Table II-23   IFF (Union Beach, NJ): Changes in Selected Wastes, 1986-1987

Chemical	Change in Waste Generation (%)	Change in Production (%)	Adjusted Change in Waste (%)	Type of Change	Reason for Change
Acetaldehyde	+26	+60	-34	Process	Self-initiated
Chromium	+12	+40	-28	Process	Disposal costs
Dichlorobenzene	+16	+20	-4	Recycle	Self-initiated
Dichloromethane	+22	+20	+2	Other	Other
Diethylphthalate	+17	+20	-3	Other	Other
Ethylbenzene	+17	+20	-3	Substitution	Disposal costs
Hydrochloric acid	+16	+20	-4	Substitution	Self-initiated
Formaldehyde	+17	+30	-13	Substitution	Self-initiated
Propylene oxide	+15	+20	-5	Other	Other
Toluene	+10	+20	-10	Substitution	Disposal costs
Xylene	+16	+20	-4	Substitution	Disposal costs

Source: New Jersey Right-to-Know Supplementary Toxic Release, Form DEQ-100, 1987.

Table II-24 summarizes information on 11 source reduction activities that were described in written materials provided by IFF to the state of New Jersey, including reductions in seven of the chemical wastes reported to the Right-to-Know program. Two chemicals, acetaldehyde and chromium, were reduced due to process modifications. Hydrochloric acid, formaldehyde, ethylbenzene, toluene, and xylene were all reported reduced through substitution of raw materials. The factor most often cited as motivating IFF to reduce its waste was reduction of treatment and disposal costs. The "other waste minimization" techniques used to reduce diethyl phthalate and propylene oxide were most likely not source reduction but such methods as on-site treatment.

The state of New Jersey requires IFF, as a condition of its air permit, to report on efforts to reduce air emissions from an aeration basin at its on-site wastewater treatment unit. In 1988, as a condition of this air permit, IFF conducted a study to reduce these air emissions. The report on this study states that the flow into the basin comes from the treatment plant's primary clarifier. In order to identify the sources of the air emissions, it was necessary to investigate the inflow to the treatment plant itself. The most frequently produced products were sampled, from at least two separate batches each. This survey identified 200 different wastestreams characterized by pH (acidity) and total organic carbon (TOC). Those wastestreams with the highest TOC were further analyzed for individual chemical components.

The specific processes generating these wastestreams were investigated by the plant process control department for ways to reduce the wastes. Through changes in reaction conditions and procedures, a solubilizer was eliminated (isopropyl alcohol) from one of the reaction steps used in the production of the plant's largest-volume product. The changes took about 4 months to implement.

For the plant's seventh largest product, the corporate research and development group discovered alternate chemistry for one of the reaction steps. Through the use of a different catalyst system and different reaction conditions, it was found that the standard reaction could be run with better yield and that the resultant process wastestream would be free of the organic chlorides that were a by-product of the original process. Implementation of this change took about 3 months.

Improved operation controls are the most common source of yield improvements at this plant. Continuing variable studies of all large-volume products at the plant are designed to allow fine tuning of the processes so that they operate as close to optimal conditions as possible. A yield improvement of 7 percent resulted from fine tuning the reaction temperature control. The improved process uses 7 percent less raw materials, and produces 7 percent less by-product. Since fewer batches are needed to produce a given volume of this intermediate, there is also a decrease in the amount of wastewater from reaction cleaning and product purification.

Another example of improved operations, also cited by IFF in its report to New Jersey on compliance with their air permit, involved the use of a better heat-up control system on a special purpose still that has resulted in a significant reduction in losses to the still vacuum system (scrubber).

The plant reports (in both 1986 and 1987 New Jersey Hazardous Waste Generator Waste Minimization Reports) that further opportunities to reduce RCRA hazardous wastes were hindered by permitting burdens, technical limitations of the production processes, and concern that product quality might decline as a result of source reduction, and that cost savings in waste management or production would not recover the capital investment.

According to IFF's New Jersey Hazardous Waste Generator Waste Minimization Reports, capital expenditures devoted to source reduction and recycling of RCRA hazardous wastes totaled $50,000 prior to 1986, and an additional $50,000 in 1986 alone. In 1987, however, there were no reported capital expenditures made for source reduction. Operating expenses for source reduction and/or recycling increased threefold from $25,000 per year prior to 1986 to $75,000 in 1987.

Technical Assistance

According to its New Jersey Hazardous Waste Generator Waste Minimization Report for 1987 source reduction technical assistance has come from within the firm or has been requested or received by IFF from other firms, consultants, and suppliers. No technical assistance has been requested or received from local, state or federal government, trade associations, or educational institutions.

Company Comments on State and Federal Regulations

According to its New Jersey Hazardous Waste Generator Waste Minimization Report for 1986, IFF would like to see government amend regulations, establish tax incentives, and provide technical assistance for source reduction.

Table II-24   IFF (Union Beach, NJ): Source Reduction Activities

Waste Medium (SR Type) Year	Source Reduction Activity	Specific Waste Reduced (Hazardous or Nonhazardous)	Percent Waste Reduced	Amount Waste Reduced
(PS) 1987	Process change.	Acetaldehyde (H)	34%	
Solid (PS) 1987	Process change.	Chromium (H)	28%	
(CH) 1987	Substitution of chemical in process.	Ethylbenzene (H)	3%	
Air (CH) 1987	Substitution of chemical in process.	Hydrochloric acid (H)	4%	
(CH) 1987	Substitution of chemical in process.	Formaldehyde (H)	13%	
(CH) 1987	Substitution of chemical in process.	Toluene (H)	10%	
(CH) 1987	Substitution of chemical in process.	Xylene (H)	1%	
Air (PS) 1988	Changes in reaction conditions and reaction procedures allowed elimination of a solubilizer from one reaction step.	Isopropyl alcohol (H)	100%	
Air (PS) 1988	Use of different catalyst system and reaction conditions improved yields and eliminated organic chlorides as by-product.	Organic chlorides (H)	100%	
Air/water (OP) 1988	Fine tuning of reaction temperature control improves yield. Fewer batches means decrease in wastewater from reaction cleaning and product purification.		7%	
Air (OP) 1988	Use of better heat-up control system on a special-purpose still results in fewer losses to still vacuum system.			

Key to source reduction types: CH, chemical substitution; EQ, equipment change; OP, operational change; PR, product change; PS, process change.
A blank indicates that the plant did not provide information.

Change in Yield	Dollars Saved	Dollars Spent	Motivation	Comments	Time Needed for Implementation
			Self-initiated review	Overall waste generation increased by 26% but production increased by 60%.	
			Disposal costs	Overall waste generation increased by 12% but production increased by 40%.	
			Self-initiated review	Overall waste generation increased by 17% but production increased by 20%.	
			Self-initiated review	Overall waste generation increased by 16% but production increased by 20%.	
			Self-initiated review	Overall waste generation increased by 17% but production increased by 30%.	
			Disposal costs	Overall waste generation increased by 10% but production increased by 20%.	
			Disposal costs	Overall waste generation increased by 16% but production increased by 20%.	
			Air permit conditions		4 mo
			Air permit conditions		3 mo
+7%			Air permit conditions		
			Air permit conditions		

MAX MARX COLOR AND CHEMICAL COMPANY
Irvington, New Jersey

Summary

Max Marx is a small manufacturer of organic pigments that employs 20 people at a plant built in 1908 in Irvington, New Jersey, west of Newark. Although the British firm Johnson Matthey bought Max Marx in 1980 (and sold it back to its original owners in 1988), it was primarily to sell Johnson Matthey products and did not affect the ongoing operations at the plant. Due to changes in its product line, Max Marx no longer uses aniline, which was the only regulated hazardous waste handled at the site.

In 1987, an operations manager was hired and, among other duties, he is responsible for environmental compliance. The major wastestream from this plant is wastewater to the municipal sewage treatment plant, which Max Marx monitors for excess color.

Max Marx granted an on-site interview to INFORM and conducted a tour of its Irvington, New Jersey, facility for this study as well as INFORM's 1985 study, but had no source reduction activities to report.

Products and Operations

Max Marx is a small speciality organic pigment manufacturer. Its products are used in printing inks and paints; its sales are less than $5 million a year. The plant has 20 employees. It uses stirred batch reactors, not being large enough for continuous operations. An on-site laboratory is used to develop products according to customer specifications, but not for research.

Operations at this site include both dry blending operations for inorganic pigments and synthesis of organic pigments, using water-based reactions and no solvents. Max Marx purchases intermediates such as beta-napthal, BONA, aromatic amines, and naphthalene sulfonic acid. The wastewater effluent is primarily salts, and is sent to the local sewage treatment plant where the small amount of pigment settles out and the effluent receives secondary treatment. Max Marx reports that it does not generate any hazardous wastes.

In the last 5 years, Max Marx has had a few product changes, although its overall volume is about the same. Due to changes in its product line, aniline is no longer in use at the plant. Due to market conditions, Max Marx no longer manufactures intermediates, so the pyrazolone process is no longer employed at the plant.

In 1980, Johnson Matthey, a British company, bought Max Marx in order to use its sales force to sell Johnson Matthey products in the United States. In 1988, Johnson Matthey sold Max Marx back to the original owner.

Environmental Policy

Max Marx's environmental policy is to abide by federal, state, and local laws. In 1987, an operations manager was hired and is responsible for environmental compliance at the plant. Prior to this, the sales manager fulfilled these duties and he has filed two premanufacture notifications (PMNs) in the last few years. The operations manager spent 8 years at large chemical companies before coming to Max Marx.

With the sale of Max Marx back to the original owner, New Jersey's Environmental Conservation and Recovery Act (ECRA) required the plant to show that the site was free from hazardous materials that might have been spilled accidentally at the site.

Materials Data Collection

Although no regulated hazardous wastes are generated during the synthesis operations at this plant, Max Marx does monitor these operations using chemical spot tests to ensure that the constituents of the reaction are being used to their maximum efficiency. The color is also checked, by visual test, at the end of a run to ensure that it is of the right quality.

Waste Management Practices Noted

Although it did not report any source reduction activities, Max Marx did describe its waste management practices. In recent years, the company has focused on its housekeeping operations. For example, when the product filter presses are washed, the rinsewater is collected in a trough and pumped to a vat where it is bleached before being sent to the sewage treatment plant. Although these are not hazardous wastes, Max Marx bleaches them because its discharge permit does not allow excessive color in the effluent.

In 1987, in samples taken by the sewage treatment plant, some heavy metals were found. These are now coagulated and collected in a drum. The plant has had the wastes tested to ensure that they are not hazardous so that they will not have to be manifested as regulated hazardous wastes.

There are some dust emissions from the organic powders. These are collected by a baghouse dust collector. The baghouse replaced a wet scrubber from which water with small amounts of pigment was sent to the sewage treatment plant. The baghouse dust is currently stored, but not enough has been collected to necessitate disposal yet. So far, there is enough storage room that disposal is not a pressing problem. This too is not regulated hazardous waste and so does not have to be manifested.

There have not been any manifested (RCRA-regulated) wastes since 1986 when Max Marx had to manifest and dispose of ten 55-gallon drums of raw materials that were never used. Any off-grade materials at Max Marx are sold to a salvage dealer for resale.

A major cost increase for the plant has been the escalating cost of garbage disposal. This affects the disposal of drums. Some of Max Marx's suppliers use recycled drums, and Max Marx uses reconditioned drums to pack its products.

Technical Assistance

The operations manager said he would find it helpful if there were someone in the state government who could answer questions for small companies.

MERCK AND COMPANY, INC.
Rahway, New Jersey

Summary

Merck and Company's corporate headquarters, research and development facilities, and a large chemical manufacturing plant, built in 1903, are located in Rahway, New Jersey. During the mid-1980s, as part of Merck's Chemical Manufacturing Division restructuring plans, this site's operations shifted from the manufacture of bulk organic chemicals to, primarily, research and development of new products.

Merck's corporate environmental policy emphasizes adhering to both the spirit and the letter of all laws and regulations and providing customers with information on how to handle Merck products in an environmentally responsible way. Specific corporate objectives include source reduction, particularly through innovative research.

To this end, and through the new focus on new product development, a team of computer scientists and development engineers has developed a computer system, called PROVAL, for use at the Rahway site. The system simulates a new batch process so that modifications to improve yield and reduce waste generation can be incorporated at the pilot scale before major capital investments are made. It produces a description of all types of wastestreams expected from the process and develops an environmental assessment based on this. So far, the system does not include air emissions and, Merck officials report, operator habits (such as whether an operator is careful to avoid spills) are also difficult to incorporate into the system.

Merck granted an interview for INFORM's 1985 study and at that time reported four source reduction activities that reduced 3,263,000 pounds of waste and saved the company $47,750 each year. Merck did not grant an on-site interview to INFORM for this study but provided information through several telephone interviews and written materials. Merck officials did not directly report any new source reduction activities to INFORM for this study; however, in 1988 New Jersey legislative testimony, Merck officials described one source reduction activity reducing 9.7 million pounds of waste and saving the company $1 million each year.

Products and Operations

Merck and Company's site in Rahway, New Jersey, houses a chemical manufacturing plant along with research and development facilities and is currently the corporation's headquarters. While overall corporate sales increased an average of 11 percent per year between 1984 and 1987, Merck's 1987 annual report stated that the company's Chemical Manufacturing Division was being restructured. The restructuring plans included phasing down the Rahway manufacturing facilities to an operation almost totally devoted to new product introduction. Previously, the Rahway plant manufactured bulk organic chemicals for use in health products and in pesticides. This is reflected in the decrease in production employees from 500 in 1984 to 200 in 1986, as reported in Merck's 1986 New Jersey Environmental Survey, Part II, a decrease of 60 percent in 2 years. However, nonproduction employment in Merck's research and corporate office facilities at the Rahway site increased from 3,500 to 3,800, or 9 percent, during the same period.

Environmental Policy

In April, 1990, Merck and Company, Inc.'s chairman and chief executive officer released "Merck and the Environment: A Commitment to Excellence," an environmental policy outline for all company employees. The policy states that Merck:

" 1. Complies with both the spirit and the letter of all laws and regulations intended to protect health and the environment;

2. Provides the same high level respect for the environment and the community worldwide as it does for its domestic facilities; and

3. Provides its customers with appropriate information to allow them to handle Merck products in an environmentally responsible manner."

Stated policy objectives are:

" 1. Minimizing [through source reduction, on- and off-site recycling, and end-of-pipe pollution control] the release of chemicals into the environment that could affect health, deplete the ozone layer or contribute to acid rain, the greenhouse effect or any other global environmental problem;

2. Seeking through research, innovative routes to waste minimization and resource conservation;

3. Minimizing the generation of wastes and seeking self-sufficiency in treating and disposing of wastes;

4. Applying in its research, manufacturing, office and vehicular fleet operations energy and resource conserving practices; and

5. Promoting resource conservation through innovative package design and the use of recyclable materials."

Materials Data Collection

In 1986, Merck introduced PROVAL, a computer-simulated batch *process eva*luation software system, at the company's Rahway, New Jersey, facility. The primary function of the system, according to a 1988 paper presented by Merck at the 43rd Industrial Waste Conference in West Lafayette, Indiana, is to maximize process efficiency, including reduction of waste, prior to full-scale construction of process operations, since changes made to processes past the pilot scale are generally more difficult and costly to implement. It took a team of Merck's corporate computer scientists and development engineers almost 12 years of multimillion dollar research and development to develop PROVAL.

The system is used during the early development stages of new products and processes at the plant so that modifications can be made prior to major capital investments. According to Merck, PROVAL is especially appropriate for the Rahway facility since this site is now primarily devoted to new product introduction. The system is used on all new products and processes at the Rahway facility.

PROVAL is designed to facilitate the evaluation of the overall impact of process changes on the productivity and economics of the process. It enables process design engineers to identify source reduction opportunities and estimation of savings achievable. The system:

- produces an energy and materials balance of a process,
- describes the kinds and amounts of chemical-specific wastes released to the air, water, and land from each step of a production process, and
- accounts for the costs associated with treatment and disposal of waste released to water and land (it will eventually account for costs associated with air emissions as well).

The PROVAL system works by using basic information supplied by the company's research chemists and engineers who are developing the new process and/or product. This information includes:

- the name of each chemical raw material used and its characteristics (such as solubility, volatility, etc.),
- the types and characteristics of by-products generated from the chemical and physical interactions between these raw materials, and
- size and efficiency of process equipment.

PROVAL simulates actual full-scale operations by varying process conditions such as temperature, pressure, reaction time, and equipment sizing. Results of the PROVAL analysis are sent back to the chemists and engineers who take a new look at the process. When the best design is deduced in terms of both economics and waste generation, a pilot-scale process operation is constructed for further testing prior to full-scale implementation.

PROVAL generates a description of multimedia wastestreams generated at the process level. The information on wastestreams is compartmentalized and quantified and then input into a second computer system now being developed. The second system is called EASY (*E*nvironmental *A*ssessment *S*ystem). EASY categorizes and ranks the wastestreams in terms of their probable fate in the environment and the best way in which to manage the waste. Economics of treatment/disposal options are also described. EASY currently can evaluate liquid and solid wastes; Merck is working on including air emissions too. The second version of the program will also be able to describe and rank waste streams in terms of toxicity.

Both PROVAL and EASY are limited by the input given to the systems by the process engineers and chemists. In terms of the programs' usefulness in identifying all source reduction opportunities, the system does not take into account either maintenance or operational practices of process technicians that may result in waste generation.

According to Merck, the major obstacle to using PROVAL in refining existing processes is that it is not yet available on a user-friendly personal computer (PC) system, and is therefore not as accessible. Merck told INFORM that there has been a "great deal" of effort to "migrate" PROVAL to the PC level.

Officials also noted that a major limitation of the PROVAL system as a means of identifying all possible sources of waste in a process is that it cannot identify opportunities for improvements in operator handling of chemicals and equipment. For example, PROVAL may show that a given process can efficiently convert 99.9 percent of the input raw materials into product, but it cannot indicate how efficient the transfer of the raw material into or out of the process may be. PROVAL cannot answer the question of whether or not the operator is careful to avoid loss of raw materials through spills or evaporation.

Merck did not provide information on waste for this study. However, data from governmental sources illustrates some waste generation patterns at the Rahway plant. Table II-25 shows changes in quantities of 10 chemicals used at and released by the Rahway facility between 1978 and 1988. For 1978 and 1985, data are available on both chemical input and chemical waste, while the Toxics Release Inventory (TRI) data used for 1987 and 1988 cover only amounts released to the environment or transferred off-site as wastes. Four chemicals reported by Merck in 1978 (chloroform, formaldehyde, phosgene, and propylene oxide) were no longer reported in 1988. However, it is not known if this is the result of source reduction efforts, changes in products manufactured at the plant, or use in quantities below the threshold for TRI reporting.

For 1978 and 1985, Table II-25 shows the releases to the environment (as air emissions, wastewater discharges, solid wastes, and off-site transfers) as a percent of the chemical input. Releases of two chemicals as a percent of input increased between 1978 and 1985, from less than 1 percent to 14 percent for phosgene and from 0 to 17 percent for propylene oxide. Releases of benzene, toluene, and tetrachloroethane decreased from 100 percent of input for each in 1978 to 92, 80, and 63 percent of input, respectively, for 1985. Releases of aniline were nearly zero as a percent of input in both years, while chloroform releases were 100 percent of input in both years. Table II-25 also shows that, from 1978 to 1988, total environmental releases decreased for four chemicals (aniline, benzene, dichlorobenzene, and tetrachloroethane) and increased for two (dichloromethane and toluene), while no 1988 data were available for four (chloroform, formaldehyde, phosgene, and propylene oxide).

Table II-26 presents the data submitted by the Merck Rahway facility on its Toxics Release Inventory (TRI) report forms for 1987. The facility released or transferred more than threshold amounts of 21 of the toxic chemicals on the TRI list. Overall TRI releases and transfers increased by over 1.3 million pounds, about 15 percent, between 1987 and 1988. Methanol accounted for most of the waste quantity of these chemicals and for the increase: 5.6 million pounds, or 64 percent of the total in 1987, and 7.0 million pounds, or 69 percent of the total, in 1988. Transfers off-site and to public sewage systems increased from 1987 to 1988, while direct releases to the air, land, and water decreased by 38.5 percent.

Table II-25 Merck (Rahway, NJ): Chemical Input and Waste Generation, 1978–1988*

Chemical	1978 Input (lb)	Waste (lb)	%	1985 Input (lb)	Waste (lb)	%	1987 Waste (lb)	1988 Waste (lb)
Aniline	850,000	0	0	1,077,660	88	<1	16,100	144,439
Benzene	1,466,000	1,466,000	100	205,850	189,509	92	262,400	241,569
Chloroform	95,000	95,000	100	16,880	16,927	100	15,403	No data
Dichlorobenzene	310,000	310,000	100	0	0		135,000	61,430
Formaldehyde	2,000	2,000	100	0	0	NA	No data	No data
Dichloromethane	136,000	136,000	100	0	0	NA	522,000	758,600
Phosgene	3,509	5	<1	2,500	357	14	No data	No data
Propylene oxide	2,000	0	0	2,244	374	17	No data	No data
Tetrachloroethane	62,000	62,000	100	36,300	22,972	63	No data	2,443
Toluene	397,000	397,000	100	90,950	72,820	80	211,150	517,085

* "*Input*" is the amount of substance brought on site, plus the amount of substance produced on site, minus the amount of substance chemically transformed (consumed) on site; "**waste**" is the amount of this substance released to all environmental media (air, water, and land) or sent off-site for disposal (after any on-site treatment); "**%**" is the percent of the substance input to the facility that is subsequently released as waste; **NA**, not applicable.

Sources: 1978 data, New Jersey Industrial Survey, 1980; 1985 data, New Jersey Environmental Survey, Part II, 1986; 1987 and 1988 data, SARA Title III, Section 313, Toxics Release Inventory (TRI) Form R.

Other Source Reduction Program Features

In addition to having a formal source reduction policy, materials balance/materials accounting, and cost accounting, Merck also has all of the four other program features tracked by INFORM: an environmental policy, environmental goals, leadership, and employee involvement. However, there is no indication that Merck has established a program to systematically address source reduction for existing products and processes at the Rahway facility as it has done for new processes and products through the PROVAL system.

Merck has established the following corporate environmental goals: (1) by the end of 1991, reduce by 90 percent worldwide air emissions of carcinogens and suspected carcinogens; (2) by the end of 1993, totally eliminate these air emissions or apply best available control technology; and (3) by the end of 1995, reduce by 90 percent worldwide all environmental releases of toxic chemicals. The basis for these emissions will be the 1987 Toxics Release Inventory "as adjusted for foreign operations."

Merck division presidents have primary responsibility for meeting these goals and assuring that programs are established to train employees to assist in doing so. The senior vice-president for engineering and technology is responsible for setting environmental goals and monitoring compliance. The vice-president for public affairs, in cooperation with line and staff executives, is responsible for communicating the company's policies and positions to public officials. The facility managers' responsibilities include source reduction, recycling, end-of-pipe pollution control, and keeping senior management aware of problems and progress.

Merck did not provide information to INFORM to clarify whether the Rahway facility has established a formal source reduction training and incentives program for employees. However, beginning in 1990, the performance evaluation of all salaried employees included information on progress made towards achievements in source reduction, recycling, and end-of-pipe pollution control. The April, 1990, message to employees from Merck's chairman and chief executive officer, "Merck and the Environment: A Commitment to Excellence," states that "If you use chemicals or other resources in your job, whether in manufacturing, the laboratory or support services, use and dispose of them appropriately. Ordering only the amounts of materials you need will avoid having to dispose of the surplus. If you see an opportunity for waste reduction, better chemical handling, process changes or any other environmental improvement, contact your site environmental staff. Every employee should consider prevention of occurrences that jeopardize safety, health and the environment a prime personal responsibility."

Table II-26 Merck (Rahway, NJ): Toxics Release Inventory Releases, 1987 and 1988 (pounds/year)

Chemical	Air Emissions	Surface Water	Land Disposal	Public Sewage	Off-Site Transfers	Total
1987						
Acetone	97,000	50	150	850,000	100,000	1,047,200
Acetonitrile	8,700	0	0	140,000	80,000	228,700
Ammonia	159,500	200	600	0	400	160,700
Aniline	12,100	0	0	2,000	2,000	16,100
Benzene	22,000	100	300	200,000	40,000	262,400
tert-Butyl alcohol	1,500	0	0	60,000	550	62,050
Carbon disulfide	11,800	10	30	1,000	7,100	19,940
Chlorine	7,700	0	400	0	200	8,300
Chlorobenzene	1,100	10	50	4,000	10,000	15,160
Chloroform	5,403	0	0	1,600	6,000	13,003
Cyclohexane	3,000	5	10	1,500	8,000	12,515
1,3-Dichlorobenzene	31,900	0	0	28,000	75,100	135,000
Dichloromethane	71,000	200	800	225,000	225,000	522,000
Ethylene glycol	3,120	0	0	58,000	100	61,220
Hydrochloric acid	161,000	20	100	0	1,000	162,120
Methanol	184,000	1,000	8,500	5,000,000	420,000	5,613,500
Methyl ethyl ketone	440	50	200	1,700	22,200	24,590
Methyl isobutyl ketone	12,000	0	0	1,500	41,200	54,700
Phosphoric acid	150	0	0	0	300	450
Sodium hydroxide	21,850	50	100	0	10,000	32,000
Sulfuric acid	80,430	0	0	0	500	80,930
Toluene	30,400	150	600	100,000	80,000	211,150
Total	926,093	1,845	11,840	6,674,300	1,129,650	8,743,728
1988						
Acetone	89,000	0	0	247,000	295,000	631,000
Acetonitrile	8,300	0	0	180,000	38,800	227,100
Ammonia	34,000	0	0	0	0	34,000
Aniline	4,380	0	0	140,059	0	144,439
Benzene	16,900	0	0	220,069	4,600	241,569
Carbon disulfide	13,900	0	0	1,000	10,000	24,900
Chlorine	4,490	0	0	0	0	4,490
Chlorobenzene	2,600	0	0	4,603	73,000	80,203
Cyclohexane	765	0	0	2,100	42,900	45,765
1,3-Dichlorobenzene	0	0	11,430	50,000	61,430	
Dichloromethane	128,000	0	0	3,600	627,000	758,600
Ethylene glycol	0	0	0	190,000	0	190,000
Hydrochloric acid	28,500	0	200	0	0	28,700
Methanol	122,000	0	0	5,680,000	1,210,000	7,012,000
Methyl ethyl ketone	1,780	0	0	10,000	14,000	25,780
Methyl isobutyl ketone	11,800	0	0	940	21,600	34,340
Sodium hydroxide	1,500	0	0	0	0	1,500
Sulfuric acid	16,200	0	0	0	0	16,200
Toluene	93,000	0	7	160,078	264,000	517,085
Tetrachloroethane	313	0	0	150	1,980	2,443
Total	577,428	0	207	6,851,029	2,652,880	10,081,544

Source: SARA Title III, Section 313, Toxics Release Inventory (TRI) Report Form R for 1987 and 1988.

Source Reduction Activities

In 1988 New Jersey legislative testimony, Merck officials reported on one source reduction activity; plant process engineers later provided INFORM with more information. This activity is summarized in Table II-27 and described below. For the 1985 *Cutting Chemical Wastes* study, Merck described four source reduction endeavors.

In testimony given at a joint New Jersey Senate and Assembly hearing in 1988, the executive director of Merck's corporate environmental resources described an equipment change planned for the Primaxin antibiotic production process at the Rahway facility. The change, Merck employees later reported to INFORM, was put into operation in early 1989, and cut in half the amount of methylene chloride solvent waste generated and released to the air and water from the process.

Formerly, a portion of the spent solvent was sent to an off-site waste management facility where it was recovered and returned to the Rahway plant. Plans to increase production of the antibiotic would have resulted in a 16-fold increase in the number of truckloads of methylene chloride waste leaving and returning to the facility. Financial risks associated with off-site recovery, ranging from potential accidents to interruptions in production scheduling, convinced management at Merck to seek ways to reduce the generation of waste solvent from the process.

At a cost of about $1 million, the company installed new equipment to reduce generation of waste solvent by 50 percent; however, the total quantity of waste solvent generation remains unchanged because production output of this antibiotic has doubled. Had this source reduction measure not been taken, the process would have generated another 1 million gallons of methylene chloride waste per year at an annual added cost of about $1 million.

The company's 1986 New Jersey Environmental Survey, Part II, report for the Rahway plant stated that its "Waste Minimization Program includes, for example, recycling, process modification research, and improved operations due to housekeeping, training, and inventory control." No further details were included. Nor did Merck provide information on source reduction in the optional section of its TRI report forms that allows the company to document achievements made in reducing waste generation.

Technical Assistance

Merck states that the corporation "provides the scientific, technical and financial resources needed to fulfill the Company's policies and goals."

Table II-27 Merck (Rahway, NJ): Source Reduction Activities

Waste Medium (SR Type) Year	Source Reduction Activity	Specific Waste Reduced (Hazardous or Nonhazardous)	Percent Waste Reduced	Amount Waste Reduced
Water/air (EQ) 1989	Installed process equipment to improve purification efficiency of Primaxin antibiotic production.	Methylene chloride (H)	50%	About 1,000,000 gal/yr (9,700,000 lb/yr)

Key to source reduction types: CH, chemical substitution; EQ, equipment change; OP, operational change; PR, product change; PS, process change.
A blank indicates that the plant did not provide information.

Change in Yield	Dollars Saved	Dollars Spent	Motivation	Comments	Time Needed for Implementation
	About $1,000,000/yr	About $1,000,000	Plans to increase Primaxin production. Costs and risks of off-site hazardous waste treatment and transport.	Primaxin production doubled while solvent waste remains about the same.	

MONSANTO COMPANY
Port Plastics Plant
Addyston, Ohio

Summary

The Monsanto Company's Polymer Plastics Division (its largest division) operates the Port Plastics plant in Addyston, Ohio, situated on the banks of the Ohio River in the industrial area east of Cincinnati. This plant, built in 1952, manufactures plastics, resins, and latexes that are used in making construction materials, automotive parts, adhesives, paints, and a wide variety of other consumer products. In 1985, a new melamine-formaldehyde resins production unit replaced a phenol-formaldehyde resins unit, which had been closed due to economics. While formalin production has ceased at this site and formalin is purchased from other Monsanto plants, other production units expanded in the period from 1983 to 1988. Overall employment and production levels have remained about the same at this large (850-employee) facility.

Monsanto has aggressive corporate guidelines that address air, water, and solid waste management. The solid waste management guidelines have specified source reduction as the highest priority since 1982. There are also corporate goals: to reduce toxic air emissions by 90 percent by 1992 and to reduce all process waste by 70 percent by 1995.

The Port Plastics plant has a "waste reduction coordinator" who monitors source reduction, waste management, and compliance with regulations and advises plant operators on these topics. Plant officials have found that source reduction is best accomplished by replacing old, outdated processes with newer ones. Monsanto requires that when plants expand, they must eliminate or offset any additional waste. Reduction and treatment options are reviewed for all major capital projects.

Monsanto granted an on-site interview to INFORM and conducted a tour of its Addyston, Ohio facility. The plant reported a total of eight source reduction activities reducing 17,329,900 pounds of waste and saving the company $3,764,100 each year. Monsanto also granted an interview for INFORM's 1985 study and at that time reported five other source reduction activities.

Products and Operations

The Port Plastics plant produces plastics, resins, and latexes that are used in making construction materials, automotive parts, adhesives, paints, and other consumer products. It polymerizes, formulates, pelletizes, and extrudes its plastic products before delivery to customers. Its major products are acrylonitrile-butadiene-styrene (ABS), styrene-acrylonitrile (SAN), polystyrene plastics, styrene-maleic anhydride (SMA) resin, and Resanines and Tomilor.

Changes at the Monsanto Company's Port Plastics plant since 1983 include the shut-down in 1984 of the phenol-formaldehyde resins unit, and its replacement in 1985 with a modernized, automated melamine-formaldehyde resins unit; shut-down of the formalin production process in the spring of 1987; construction/start-up of a coal-fired boiler in the fall of 1986; and several major expansion/modernization projects in the ABS area. The phenol resin business has been discontinued because of economics. The replacement process makes different products for different customers. Formalin production has been eliminated because sufficient stock can be purchased from other Monsanto units.

Environmental Policy

The Monsanto Company established a set of worldwide environmental guidelines in the late 1970s and early 1980s that serve as guidance for internal environmental programs throughout the corporation. In 1986, Monsanto's chairman of the board appointed an Environmental, Safety and Health Committee (based on a prior Environmental Policy Committee) to ensure that

Monsanto develops appropriate policies, guidelines, and procedures in the environmental, safety, and health areas and that operating units perform in a manner consistent with these policies and guidelines. Representatives on the committee come from all operating units and subsidiaries as well as from the corporate environmental, health, and safety group. The committee reports to Monsanto's senior vice-president for environmental safety and health.

Monsanto's ultimate goal is zero effects from discharges worldwide, with a specific goal established to reduce air emissions of TRI chemicals by 90 percent by the end of 1992. Additionally, Monsanto has set a goal to reduce process wastes by 70 percent by the end of 1995. A hierarchy of waste management options has also been established to ensure that the best environmental alternative is chosen for wastes that remain.

Monsanto's first worldwide guideline addresses air and water programs and was intended to anticipate future regulations. In 1987, under this guideline, and in anticipation of regulations on hazardous air emissions, the Port Plastics plant sampled 70 air vents and estimated air pollutant health and safety factors with a generalized diffusion model that models air dispersing in the environment downwind from the source of emissions.

The second worldwide guideline addresses hazardous solid wastes. The hierarchy of management options, as stated in this guideline is: (1) source reduction, initiated in 1982; (2) reuse, recycling, or co-product sale; (3) incineration or other treatment to make waste less hazardous; and (4) land disposal of treatment residues or wastes not amenable to higher order management options. The guidelines also specify corporate programs, updated periodically, which address both source reduction and other waste management options.

Since 1982, annual reduction goals have been set and reported through the corporate Manufacturing Management Council which was established in the mid-1970s. The goals are stated as a normalized percentage reduction per pound of product. Due to reorganizations in the corporate structure, no goals were set in 1987; they resumed in 1988. Progress in reduction of waste is measured and tracked against the goal of a 70 percent reduction in process wastes for Monsanto.

Each of Monsanto's operating divisions and manufacturing locations assigns a "waste reduction coordinator" who is responsible for monitoring and facilitating that unit's reduction efforts for all types of wastes. The coordinator is also responsible for waste management and compliance with regulations. Any recommendations are implemented through process and project engineering groups and various teams.

Materials Data Collection

Monsanto first compiled solid waste inventories in 1982 to provide base information for the start of the corporate waste management program, and there are annual updates to the corporate waste database. Monsanto reported that the updates take 2 months to fill out each year due to complexities introduced by the Toxics Release Inventory (TRI) legislation. This database gives the pounds of waste per pound of product for each solid wastestream. These data are fed into the Chemical Manufacturers Association database on chemical industry RCRA waste generation, which is published annually (it does not include wastewater or air emissions).

The Port Plastics plant has a full cost accounting system for hazardous solid waste disposal costs, including fuel credits for those units that contribute spent monomers for energy recovery. Each production department is also allocated wastewater treatment and solid waste disposal costs. Hazardous solid waste disposal shows up as a line item; wastewater and nonhazardous solid wastes are allocated on a historical basis, based on the preceding year's experience and not the current year's amounts. Any air pollution control equipment costs are allocated back to the production unit.

Other Source Reduction Program Features

In addition to having a formal source reduction policy, materials accounting (but not materials balance), and cost accounting, Monsanto's Port Plastics plant has the other four source reduction

Table II-28 Monsanto (Addyston, OH): Source Reduction Activities

Waste Medium (SR Type) Year	Source Reduction Activity	Specific Waste Reduced (Hazardous or Nonhazardous)	Percent Waste Reduced	Amount Waste Reduced
Solid (PS, PR) 1984-1985	The phenol-formaldehyde resins unit was replaced by a methylated melamine-formaldehyde resins process.	Phenol-formaldehyde resin (H)	89%	16 drums/mo average (28,800 lb/yr)
Water (EQ) 1985	The new process (above) also has a methanol recovery distillation column that enables methanol in the wastestreams to be closed-loop recovered.	Methanol (H)	100%	15,600,000 lb/yr
Solid (OP) 1985	Improved operations prevents build-up of paraform wastes in tanks.	Paraform (H)	98%	103,000 lb over 2 yr
Water (PS) 1987	The plant has upgraded the Scripset resin filter press to reduce leaks from the press.	Scripset press leakage (H)	96%	52 drums/mo (249,600 lb/yr)
Water (PS) 1986-1987	Statistical process control measures were applied to the ABS process to chart key indicators, including solids in wastewater.	Polymer solids		
Solid (PS) 1972-1987	Process changes to enable additional recovery and sale of spent monomers.	Monomers (H)	50%	1,400,000 lb/yr
Solid (PS) 1986	The ABS/SAN compounding process was modified to reduce off-grade product.	ABS plastics		
Solid (OP) 1986-1988	The plant has paved and isolated the area under each storage tank in order to keep spills from contaminating the ground.	Spills (H, N)		

Key to source reduction types: CH, chemical substitution; EQ, equipment change; OP, operational change; PR, product change; PS, process change.
A blank indicates that the plant did not provide information.

Change in Yield	Dollars Saved	Dollars Spent	Motivation	Comments	Time Needed for Implementation
	$4,800/mo		Market demand for new products. Older processes cannot compete on a production cost basis.	The new resins process is much cleaner, generating only two drums of hazardous waste per month.	
	$1,560,000/yr		Market demand.	Savings are for replacement material costs. Waste is not generated due to process design.	
	$96,500 over 2 yr		Reduce waste generation and disposal costs.		
	$300,000/yr	$60,000	Process optimization.	Savings are from avoided incineration costs.	None
+11.5%	$500,000 to $1,000,000/yr	Negligible	Improved yields to reduce costs.		Less than 1 yr
			Continuous effort to upgrade processes.		20 yr and ongoing
+5%	$1,000,000/yr		Continuous effort to improve yields and reduce costs.		
		$500,000	Environmental protection and reduction in clean-up costs.	If a spill does occur, the volume of contaminated soil to be cleaned up is not large and some spill material may be salvageable.	

program features tracked by INFORM: an environmental program, environmental goals, leadership, and employee involvement.

The Port Plastics plant has a written "hazardous and solid waste minimization" plan that includes source reduction efforts for RCRA hazardous (subtitle C) and nonhazardous (subtitle D) waste. Reduction of water effluents and air emissions is carried out in accordance with Monsanto's worldwide guidelines.

At the Port Plastics Plant, Monsanto has found that source reduction could best be implemented by process replacement; that is, by replacing an older process with the next generation designed process. This has been done primarily because of market demand for new products and because older processes could not compete on a production cost basis. The second most effective source reduction technique at the plant is optimization of existing processes; that is, improving yield and reducing costs through upgrading process controls and practices, accomplished by means of numerous short-term and long-term engineering projects. Improving the purity of materials previously disposed of as wastes so they can be reused or sold represents another means of reducing waste, especially when a waste material can become a feed material for another product.

Monsanto has a corporate cost reduction program that includes cost reduction goals and publishes examples of cost reduction projects. The Port Plastics plant manager reported that this corporate program has also been helpful in identifying source reduction measures and has found the published case histories from other Monsanto plants to be the best way of learning about ways to reduce waste in areas where it is hard to measure exactly where the waste is coming from (for example, at the wastewater treatment plant where so many wastestreams are combined).

The plant manager also cited Monsanto's corporate quality control program as an effective tool for learning about source reduction measures. This program includes statistical process control (SPC) measures and emphasizes ways to analyze problems step-by-step. Case histories are published in monthly reports. Each year there is a contest among Monsanto's plants for the best quality control measures with a Conference of Champions in St. Louis where the six best are chosen. The SPC measures were first used at the Port Plastics plant in 1986.

Within the corporation, all new major capital projects undergo thorough review at the corporate level, including a "loss prevention and environmental control" review, before approval. At these preliminary stages, the options for hazardous and nonhazardous solid waste reduction, air emissions and control, and wastewater effluents and treatment are reviewed. Plants that expand must eliminate or offset any additional waste.

Source Reduction Activities

Monsanto described the implementation of eight source reduction activities at the Port Plastics plant to INFORM for this report; they are summarized in Table II-28 and discussed below. For the 1985 report, *Cutting Chemical Wastes,* the plant had reported five other source reduction activities.

The need to develop economically competitive production processes motivated two source reduction activities. For one, a process replacement, Monsanto reported that the shut-down of the phenol-formaldehyde resins unit eliminated an average of 16 drums of hazardous waste per month and saves Monsanto $4,800 per month in disposal costs. The process was replaced by a methylated melamine-formaldehyde resins process, that generates only two drums of hazardous waste per month.

In a related equipment change, a methanol recovery distillation column was designed into and constructed along with the new process. It enables methanol in the wastestreams to be recovered on a closed-loop basis. By recovering and not generating the waste, Monsanto estimates that it annually saves $1,560,000 in costs that would have been incurred for replacement of 15,600,000 pounds of methanol. The process replacement took place from 1984 to 1985.

To reduce waste disposal costs, improved operations at the Port Plastic plant since 1985 have prevented build-up of paraform (polymerized formaldehyde waste) in tanks. The amount of waste not generated was 103,000 pounds over 2 years, for a 98 percent reduction and a savings of $96,500.

Another source reduction activity that reduced disposal costs, in this case hazardous waste incineration costs, involved upgrading the Scripset resin filter press to reduce leaks from the press. This process optimization reduced waste generation from press leakage from 54 to 2 drums per month (or 96 percent), with savings of $300,000 in 1988. The upgrading cost $60,000.

Statistical process control measures were applied to increase polymer recovery yields in an ABS polymerization process by 11.5 percent; key indicators, including solids in wastewater, were charted. Savings range from $500,000 to $1 million per year, depending on product demand, and the solid losses to the wastewater treatment plant have been reduced. Implementation costs were negligible. The application of the SPC measures took place from 1986 to 1987, and the research to develop the measurement process took less than one year.

In-process recycling of spent monomers was built into polymer production lines at the time of construction in 1972. From then through 1987, process improvements have allowed an additional 1.4 million pounds of spent monomers from polymerization processes to be recycled and reused in the processes. These changes are a result of ongoing research to improve yields.

In another effort to improve yields and reduce costs, a variety of unspecified types of projects increased gross yields of the ABS/SAN compounding process, which adds colorants and additives to plastics to make extruded pellets for sale, by 5 percent in 1986. This reduced off-grade production of ABS plastics. The improved yields result in a $1 million savings per year.

Finally, in a waste and spill prevention project, from 1986 to 1988, the Port Plastics plant paved and isolated the area under each storage tank at the site in order to keep spills from contaminating the ground. The costs of paving and isolating the tanks was $500,000. Now, if a spill does occur, there will not be a large volume of contaminated soil requiring disposal and some spill material may be salvageable.

Other Waste Management Practices

In addition to the source reduction activities it reported to INFORM, Monsanto described other waste management measures at the Port Plastics plant.

The "tube in the sky" is the plant's primary means of air pollution control. This is an elevated duct, 44 inches in diameter, that collects emissions throughout the plant site and carries them to the main boiler, where they are burned as fuel.

Recycling and reuse measures implemented at the Port Plastics plant included accumulating spent alumina, used to filter impurities out of raw materials, and sending it to a company that regenerates the alumina by burning off the raw material impurities. The alumina is utilized as an abrasive. This avoids the need to landfill or incinerate the spent alumina.

Shutting down the formalin production unit reduced formalin inventories on-site by about 50 percent, since half of the output was sold to other plants and manufacturers and half was used on-site. As a result, associated paraform wastes decreased from 127,000 pounds in 1984 to 103,000 pounds in 1985, 24,000 pounds in 1986, 10,000 pounds in 1987, and 3,000 in 1988. The remaining paraform wastes are due to formaldehyde used in the new methylated melamine-formaldehyde resins process.

Still bottoms from the new methylated melamine-formaldehyde process (nonhazardous waste) are burned in the plant's oil-fired boiler and the unit receives credit against its fuel costs of $1,300 to $1,400 per month.

Technical Assistance

Monsanto reported that because the Port Plastics plant is unique in many respects, including its environmental management program, assistance available from state or federal agencies is not as useful as it might be in plants in more uniform industries such as electroplating and metal manufacturing. However, the plant does use technical assistance from the corporate engineering department, particularly for new projects.

Employees of the Port Plastics plant provide technical assistance to other companies. In 1987, the senior environmental specialist at the Port Plastics plant (who is also the waste reduction coordinator) gave talks about the plant's programs to the Chamber of Commerce in Cincinnati and the Air Pollution Control Association. He has also given detailed case histories of source reduction examples at this plant at a conference attended by government, industry, and environmental groups. Environmental presentations continue with the present staff.

Company Comments on State and Federal Regulations

It is Monsanto's belief that waste generation and reduction will eventually be regulated in addition to waste management, as some states are already attempting to do. Monsanto would prefer to see incentives for source reduction rather than mandatory regulations. Officials noted that moving to the next generation of processes (for example, from the phenol-formaldehyde resins unit to the melamine-formaldehyde resins unit) can produce significant advantages in yield and, consequently, in source reduction. They believe that the capital required to replace or modernize such production facilities should be given special tax incentives, especially when source reduction results can be documented. However, they point out, the US Congress has eliminated tax deductions for this type of large capital investment, making them more and more difficult to implement.

Future

Monsanto's plans include reducing the generation of wastewater through closed-loop reclaiming and recycling of wastestreams into production processes as processes are replaced, retired, and upgraded to new processes. In addition to wastewater reduction, the modernization of the ABS polymerization processes is expected to realize improved yields that will reduce the generation of solid wastes.

Monsanto has also been studying ways to reclaim materials from its waste treatment plant solids. However, the standards of its quality control program include certifying that any outside company to which it would send solids for reclaiming must have worker safety and environmental standards that are as high as Monsanto's. So far, this has prevented Monsanto from proceeding with this project. Monsanto is currently studying methods to transform this sludge into a usable low-grade plastic on site.

MORTON INTERNATIONAL, INC.
Speciality Chemicals Group
Industrial Chemicals and Additives (formerly Carstab Division)
Cincinnati, Ohio

Summary

Morton International's plant in Cincinnati, Ohio, built in 1949, manufactures performance chemicals (chemicals that enhance the intended function of the product) for the plastics, petroleum, and road paving industries. Morton International's three main business areas are chemicals, aerospace products, and salt. This plant belongs to the corporation's Speciality Chemicals Division.

As a member of the Chemical Manufacturers Association, Morton subscribes to the 1990 Responsible Care initiative, which includes source reduction as a waste management strategy. Written reports from the plant indicate that source reduction to reduce waste disposal costs is a part of its cost reduction activities.

The plant has reported an overall reduction in the amount of waste released to the air or sent off-site or to the municipal sewage treatment plant. It is not known, however, if this is due to on-site treatment or other waste management changes rather than source reduction, since no source reduction activities were described in the information provided by Morton. There is no indication that source reduction measures have been taken at the company's Cincinnati, Ohio facility.

Morton International did not grant an on-site interview to INFORM but provided some written materials. The company also did not grant an interview for INFORM's 1985 study.

Products and Operations

This plant, established in 1949, was owned by Cincinnati Milacron, Inc. until 1980, when it was purchased by Morton Thiokol. Formerly part of the Carstab Division of Morton Thiokol, this facility is now part of the Industrial Chemicals and Additives business segment of the Speciality Chemicals Group of Morton International. The division produces performance chemicals, including organotin polyvinyl chloride (PVC) heat stabilizers, antioxidants, synthetic lubricants, asphalt additives, phosphonium salt polymerization catalysts, and extreme pressure lubricant additives.

Environmental Policy

The company is a member of the Chemical Manufacturers Association and therefore has adopted the 1990 Responsible Care initiative, which includes a "Waste Reduction Code of Management Practices." In 1990, the company claimed that because waste disposal is an added cost to product manufacture, it is constantly looking for ways to reduce waste generation to improve their competitive edge.

Materials Data Collection

Since April, 1985, this site has not conducted activities requiring a RCRA hazardous waste permit. The site now maintains the status of a generator that stores hazardous waste on-site for less than 90 days. Because the plant did not provide INFORM with data on its generation of waste, such data from governmental sources are presented in the following tables.

Table II-29 indicates that the quantity of RCRA hazardous waste sent off-site has declined 62 percent despite an overall increase in production of 29 percent. This information by itself does not indicate if this reduction is due to releases shifted to environmental media in which the chemical is not regulated, reclassification of wastestreams, on-site treatment of the wastestream, or actual reductions at source.

Table II-29 Morton International (Cincinnati, OH): RCRA Hazardous Waste, 1984-1989

Year	Waste Generation (lb)	Change from Previous Year (%)	Change in Production (%)	Adjusted Change* (%)
1984	890,000			
1985	502,000	−44%	−5%	−41%
1986	708,400	+41%	+24%	−14%
1987	594,000	−16%	+9%	−23%
1988	400,400	−33%	−5%	−29%
1989	340,400	−15%	+7%	−20%
Total		−62%	+29%	−71%

* Percent change in RCRA wastes generated after adjusting for changes in production output.

Source: Hazardous Waste Generator Annual Report, Division of Hazardous Waste Management, Ohio EPA, and Morton International, Inc.

Table II-30 Morton International (Cincinnati, OH): TRI Toxic Chemical Releases, 1987-1989

Chemical	Air Emissions (lb/yr)	Public Sewage (lb/yr)	Off-Site Transfers (lb/yr)	Total* (lb/yr)
1987				
Chlorine	1 - 499	0	0	250
Chloromethane	89,000	1,000	0	90,000
Methanol	7,600	49,000	96,000	152,600
Sodium hydroxide	0	0	3,500	3,500
Ammonia	NR	NR	NR	NR
Total*	96,850	50,000	99,500	246,350
1988				
Chlorine	1 - 499	0	0	250
Chloromethane	73,000	1,000	0	74,000
Methanol	5,700	49,000	0†	54,700
Sodium hydroxide	1 - 499	0	4,400	4,650
Ammonia	500 - 999	0	0	750
Glycol ethers	NR	NR	NR	NR
Total*	79,950	50,000	4,400	134,350
1989				
Chlorine	1 - 499	0	0	250
Chloromethane	64,000	1,000	0	65,000
Methanol	3,800	34,000	0†	37,800
Sodium hydroxide	Removed from list by US EPA			
Ammonia	1 - 499	757,000‡	0	757,250
Glycol ethers	0	0	0	0
Total*	68,300	792,000‡	0	860,300

* If ranges were reported, the midpoint was assumed, i.e., for the range 1 - 499, the amount added in the total is 250 pounds, and for the range 500-999, the amount added in the total is 750 pounds.
† Material now used in fuel blending.
‡ Ammonium salts are used for neutralization at this plant. Before 1989, no TRI reports on this chemical were required by the US EPA. For 1989, they were included in the category of ammonia.

Key: NR, not reportable (if a chemical is manufactured in amounts less than the threshold (75,000 pounds for 1987, 50,000 pounds for 1988 and 25,000 pounds for 1989), or if it is used in amounts less than 10,000 pounds, no report to TRI is required).

Sources: 1987 and 1988, SARA Title III, Section 313, Toxics Release Inventory (TRI) Report Form R; 1989, Morton International, Inc.

Table II-30 shows the releases of chemicals reported to the Toxics Release Inventory (TRI) by Morton in accordance with Section 313 of the Superfund Amendments and Reauthorization Act (SARA). This information is based on specific chemicals used or manufactured at an industrial facility and is broken down differently from RCRA wastes; however, the category of "Off-Site Transfers" in this table could include some RCRA wastes.

For TRI, releases and transfers of specified chemicals to all environmental media must be reported. Thus, for these wastes it is possible to determine whether wastes have been shifted to environmental media in which the chemical is not regulated. Table II-29 shows that, for the three chemicals (chlorine, chloromethane, and methanol) that were reported in all 3 years, off-site transfers have been eliminated. Methanol is now used in fuel blending instead of being transferred off-site for disposal. Both air emissions and discharges to public sewage systems have been reduced by 30 percent.

Source Reduction Program/Activities

Each TRI report form contains an optional section that allows the company to document achievements made in reducing waste generation. No report forms submitted by Morton contain information in this section.

Other Waste Management Practices

In explanation of the reductions achieved in both RCRA hazardous wastes and in the TRI releases, Morton has supplied information to INFORM on recycling projects undertaken at this facility. Analysis of one wastestream led to process modifications enabling the material to be used as a raw material, while another is now largely recycled. The reductions were reported to be achieved by improving recovery efficiencies. Further reductions are expected when additional equipment (at a cost of about $600,000) is installed.

While Morton did not indicate which chemicals the recycling project applied to, they did report that the methanol wastestream was now being burned as fuel. Also, on-site treatment is reported to TRI only for the chlorine wastestream.

PERSTORP POLYOLS, INC.
Toledo, Ohio

Summary

Perstorp AB, a Swedish company, operates a small-sized facility in Toledo, Ohio, built in 1971 and employing 37 workers. The plant manufactures pentaerythritol (used in making paints, inks, and synthetic lubricants) and sodium formate (used by the leather, textile, paper, and chemical industries). In 1989, a new trimethylolpropane plant opened at the site. Trimethylolpropane is used as a raw material in other processes at the site and is sold for use in industrial coatings and lubricants and as a stabilizer for plastics.

Many Swedish environmental regulations are stricter than those in the United States and this plant is held to the stricter standards by its parent company. Perstorp's environmental policy of 1988 calls for optimal environmental protection and covers all types of waste. The parent company also provides research and development and funds for source reduction projects. The Toledo plant depends on constant monitoring of all points of waste generation to ensure processes are being operated efficiently.

Perstorp Polyols is a small plant with only a few of the source reduction program features tracked by INFORM. Nevertheless, it has been able to achieve savings in both waste and costs through source reduction activities. Perstorp granted an on-site interview to INFORM and conducted a tour of its Toledo facility for this study. The plant reported a total of four source reduction activities, saving the company $40,000 each year. Perstorp also granted an interview for INFORM's 1985 study and at that time reported three other source reduction activities.

Products and Operations

The Perstorp Polyols plant in Toledo, Ohio, is one of seven plants in the United States owned by the major Swedish chemical company, Perstorp AB. Using a continuous manufacturing process, Perstorp Polyols manufactures pentaerythritol and sodium formate from formaldehyde and acetaldehyde. Pentaerythritol is used in the manufacture of paints, printing inks, and synthetic lubricants. Sodium formate is used in the leather, textile, paper, and chemical industries.

Perstorp built a new trimethylolpropane facility at this Toledo plant in 1989. The production of trimethylolpropane uses the same types of raw materials already in use at the plant. It is used in the manufacture of industrial coatings and lubricants and as a stabilizer for plastics.

Production at Perstorp's Toledo plant has increased by 10 to 15 percent since 1985. The number of employees is about the same.

Environmental Policy

As a major company in Sweden, Perstorp's environmental practices are scrutinized by the Swedish government. Also, many environmental regulations are stricter in that country than in the United States and the Perstorp Polyols plant in Toledo is held to the practices developed by its parent company to meet the stricter standards. For example, Perstorp AB is located in a small village in Sweden where water supplies are limited and no wastewater discharges are allowed. Consequently, it has developed many closed processes that are also used at the Perstorp Polyols plant in Toledo. Also, spills are watched for and it is company policy to close down operations immediately in the case of a leak.

Perstorp's written corporate environmental policy, established in 1988, calls for optimal environmental protection at all production plants. Emission levels should meet statutory limitations by broad margins, and the company should stay abreast of all new technological developments related to environmental conservation. The policy covers solid wastes, wastewater, and air emissions.

Given the small size of the Perstorp Polyols plant, its managers believe that there is no need for a formal written environmental policy beyond the corporate one. Each employee is trained in safety practices and the handling of hazardous chemicals, and all operations personnel are trained to do all the different operations jobs at the plant. The plant manager is constantly in the plant, monitoring their work.

The parent company, Perstorp AB, has a separate environmental department whose staff are available as consultants. It also provides research and engineering assistance to all its plants. The Perstorp Polyols plant in Toledo reported that the parent company provides timely approval for funds for source reduction and other environmental projects.

Materials Data Collection

Cost accounting at this plant includes the cost of wastewater discharges to the municipal sewer for each process unit. The wastewater costs are based on chemical oxygen demand (COD). The plant has not conducted separate environmental reviews to track waste materials back to their source.

Other Source Reduction Program Features

In addition to cost accounting, Perstorp Polyols has two other source reduction program features tracked by INFORM: leadership and employee involvement. It does not have a formal source reduction policy, materials accounting/materials balance, an environmental program, or environmental goals.

Employees are trained (and retrained annually) according to OSHA regulations to handle formaldehyde and other chemicals in a way that reduces releases. The plant manager, whose background is in chemical engineering, has primary responsibility for source reduction at the plant. She constantly monitors the work of operations personnel, including all points of waste generation.

Source Reduction Activities

Perstorp Polyols reported four source reduction activities to INFORM for this report; these are summarized in Table II-31 and discussed below. Perstorp described three other activities to INFORM for its 1985 report.

Perstorp has an automatic system for monitoring COD in its wastewater streams; in 1987, the company initiated analysis of COD in the wastewater every 3 minutes, 24 hours a day. This enabled operators to quickly see if a problem arises. Primarily installed to better control processes and improve yields, the system has also reduced the COD per unit of product by 30 percent; total COD in the wastewater has decreased even with an increase in production. The overall decrease in sewer charges due to reduced COD levels has been $40,000 per year.

New state VOC regulations for volatile organic compounds (VOCs), passed in 1986, required monitoring of 106 points throughout the plant. While it had always been company policy to close down if there was a leak, the plant manager reported that the new regulations have made all the operators more aware of the importance of maintaining the possible leak sources so that fewer leaks occur.

Since 1985, changes made to the sodium formate process have reduced the amount of fine particles generated. This has improved process yields and also generates less dust.

Perstorp also has improved its housekeeping procedures. Instead of washing away product spills, the spills are swept up and reused when possible.

Other Waste Management Practices

In addition to its source reduction activities, Perstorp reported some other waste management practices. For instance, in 1985, Perstorp installed a neutralization system for its wastewaters.

Table II-31 Perstorp Polyols (Toledo, OH): Source Reduction Activities

Waste Medium (SR Type) Year	Source Reduction Activity	Specific Waste Reduced (Hazardous or Nonhazardous)	Percent Waste Reduced	Amount Waste Reduced
Water (OP) 1987	Wastewater streams are automatically sampled for chemical oxygen demand (COD) every 3 minutes, 24 hours a day, to identify and correct problems as soon as they occur.	COD (N)	30% per unit of product	
Air (OP) 1986	Volatile organic compounds (VOCs) are monitored at 106 points throughout the plant.	VOCs (H)		
Air (PS) 1988	Process changes in the sodium formate process reduced amount of fine particles generated.	Dust (N)		
Water (OP) 1985	Improved housekeeping: spills are swept up instead of rinsed away, and product is reused when possible.	Product (N)		

Key to source reduction types: CH, chemical substitution; EQ, equipment change; OP, operational change; PR, product change; PS, process change.
A blank indicates that the plant did not provide information.

Previously, it relied on mixing process wastewaters to attain the pH required by the sewage treatment plant standards, but found that this was hard to do without excess amounts of caustic. As the price of caustic has increased, costs are reduced with the more precise neutralization system.

Perstorp has also installed a wet scrubber on its sodium formate process unit in addition to a cyclone that was already in place. Emissions have been reduced and, in addition, Perstorp is now able to sell the liquid from the wet scrubber as a product for use in nonphosphate detergents.

The new trimethylolpropane facility has a catalytic incinerator on all storage tanks of volatile organics.

Technical Assistance

Perstorp's environmental permits (air and water; the plant has no RCRA wastes) are issued by the Toledo Environmental Services Agency (TESA). Perstorp officials state that TESA is helpful in the interpretation of regulations and that this is especially useful since Perstorp is a small company.

The Formaldehyde Institute (a trade organization), and Du Pont (Perstorp's supplier) have helped Perstorp on issues concerning the use and handling of formaldehyde. In particular, they have provided Perstorp with videotapes for employee training. Suppliers have also given helpful assistance in employee training for other chemicals used at the plant.

Company Comments on State and Federal Regulations

The Perstorp Polyols Toledo plant expects to have no trouble meeting new OSHA standards for formaldehyde in the workplace (1 ppm instead of 3 ppm) because the lower standard was already

Change in Yield	Dollars Saved	Dollars Spent	Motivation	Comments	Time Needed for Implementation
	$40,000/yr		To improve yields.	Total COD in the wastewater has decreased even with an increase in production.	
			Regulations require monitoring.	Monitoring has made operators more aware of importance of maintenance to avoid leaks.	
			Improve yields and generate less dust.		

in effect in Sweden. It will continue monitoring and education of its employees as required by the regulation.

Recently promulgated EPA pretreatment standards for organic chemical plants do not affect this plant because there are no metals in the wastewater. The surcharge for "high strength dischargers" (as measured by the COD in the wastewater) has, however, been an incentive to reduce wastes. Perstorp has found that it has been able to reduce COD and save both the sewer charges and product through frequent analyses of its automatic monitoring data.

Future

Industrial workplace regulations on formaldehyde may get stricter in the future and require more monitoring and record-keeping.

PMC SPECIALITIES GROUP
Division of PMC, Inc.
(formerly Sherwin-Williams Company)
Cincinnati, Ohio

Summary

Established in 1966, the PMC facility in Cincinnati, Ohio, produces many kinds of organic chemicals, including saccharin, corrosion inhibitors, intermediates and additives, and speciality chemicals. PMC's plant in Cincinnati is the only US manufacturer of saccharin, isatoic anhydride, and anthranilic acid (a tarnish inhibitor). When Diet Coca Cola switched its artificial sweetener from saccharin to NutraSweet in 1985, this plant lost a major saccharin customer. PMC bought this facility from Sherwin-Williams in 1985. The plant manager has been there, under both ownerships, since 1984.

PMC is a large, privately owned chemical company with no formal environmental policy. The Cincinnati plant has a hazardous waste management policy covering RCRA waste, as recommended by the state, following a study of the plant in 1985. This PMC plant has an Environmental Group that monitors regulatory compliance and also participates in product development. During the product development process, costs of waste handling are assigned, and a list of chemicals that PMC does not want to use at the plant is checked.

Because of strict limits on the discharge of trichlorobenzene (TCB) in wastewater, PMC totally eliminated use of this chemical.

PMC granted an on-site interview to INFORM and conducted a tour of its Cincinnati facility. The plant reported a total of six source reduction activities, reducing 90,510 pounds of waste and saving the company $260,000 each year. The plant's former owner, Sherwin-Williams, did not grant an interview for INFORM's 1985 study, but information from public sources showed that two other source reduction measures had been taken at that time.

Products and Operations

PMC is the only US manufacturer of three products: saccharin (an artificial sweetener used in dietetic foods, beverages, snacks, toothpastes, mouthwashes, cosmetics, pharmaceuticals, and tobacco), isatoic anhydride, and anthranilic acid. It faces competition from international chemical companies, including ones in Germany, Japan, Korea, and China.

While some areas of production have increased and others decreased, employment and overall dollar production are about the same as in 1982, when the plant employed about 200 people.

Environmental Policy

The PMC Specialities Group, formerly owned by Sherwin-Williams, is a division of PMC, Inc., a privately owned chemical company that has no formal overall environmental policy. However, it does have a written policy on hazardous waste management. Elements of this policy include documentation of wastes generated, key contact people at the plant, and procedures for the handling and reporting of spills. The formal policy was written in response to a 1985 investigation by the Ohio Environmental Protection Agency (Ohio EPA), which found the plant in compliance with RCRA regulations but weak in the area of documentation of wastes and procedures. This policy covers RCRA wastes only.

PMC has an Environmental Group within the plant. The group's major function is to ensure that the plant remains in compliance with regulatory requirements. It also explores and completes operational improvements that relate to environmental matters. A technical manager controls the overall environmental programs, and the environmental engineer relates to regulatory agencies and resolves associated problems or concerns. The environmental engineer has been in her position since 1982, 2 years after this plant developed the position. Since 1980, reduction of wastes (air, wastewater, and solid wastes) has been one of the environmental engineer's responsibilities. As with any project at this plant, the company's policy is to pursue environment-related projects based on their technical merits and its costs.

Materials Data Collection

In 1981, an increased sewer surcharge led the company to survey where the wastes in the plant's wastewater were coming from. The survey showed that purchased materials and materials produced were not all going into product, but that some wastes were going into the sewer. In response to this, the company's Chemical Development Group instituted a study to track waste by product so that costs could be assigned. The survey has been repeated annually. Costs are updated as processes are improved. Sewer charges and surcharges are allocated according to water use in each building. In addition, a weekly inventory now provides data on raw material age and use, and product yields. In the past, a few raw materials had to be incinerated because they had been allowed to go out of date.

Other Source Reduction Program Features

In addition to having materials accounting and cost accounting (but not a written source reduction policy or materials balance, PMC reported having three of the other four program features tracked by INFORM: leadership, employee involvement, and an environmental program (but not specific environmental goals).

PMC's Environmental Group participates in chemical and product development meetings that focus attention at the beginning of product development on the types, quantities, and handling requirements of materials involved, and the costs of wastes. Costs of waste handling are assigned during the development process. There is a list of chemicals, particularly solvents and particularly relating to RCRA wastes and pollutants in wastewater, that PMC does not want to use at the plant. This list is referred to during product development, and at other management meetings.

The plant manager reported that such cooperation between PMC's chemists and environmental engineers further helps to avoid wasteful practices throughout the plant. For example, if production output is measured below normal levels in the weekly inventory, an investigation by the group determines whether the cause is in the area of equipment, chemistry, or staffing.

In addition, the plant has a "Merit Award" program that provides a monetary reward for good ideas. Some awards have been for ways to reduce waste.

Source Reduction Activities

According to PMC, six more source reduction activities have been implemented since the two activities reported in INFORM's 1985 study, *Cutting Chemical Wastes*. These six are summarized in Table II-32 and described below.

The plant manager reported that tremendous pressure from overseas competitors has forced PMC to look very carefully at losses due to waste generation. The vast majority of PMC's wastes are waterborne. Hazardous solid wastes are shipped only every 4 or 5 months, so that disposal costs are not great.

A process change that was instituted in 1987, after 6 months of research, led to the reduction of trichlorobenzene (TCB) in wastewater. TCB had previously been washed out of the product with naphtha, which resulted in an inseparable solution. PMC was able to rework this process so that the TCB could be closed-loop recovered. The waste had value but could not be used in its existing form. Modifications were needed to prevent the generation of large quantities of mixed solvent, and to recover the individual solvents. The waste was reduced by 50 percent, or 5,000 pounds per year. The cost of the change was $25,000.

Ever stricter limitations on TCB discharges to the municipal treatment plant led to further

Table II-32 PMC Specialities Group (Cincinnati, OH): Source Reduction Activities

Waste Medium (SR Type) Year	Source Reduction Activity	Specific Waste Reduced (Hazardous or Nonhazardous)	Percent Waste Reduced	Amount Waste Reduced
Water (PS) 1987	Trichlorobenzene (TCB) used to be washed out of the product with naphtha. The process was reworked so that the individual solvents can be recovered.	TCB (H)	50%	5,000 lb/yr
Water (PS) 1990	Changes in process chemistry eliminated the need for TCB.	TCB (H)	100%	
Water (OP) 1985	A discharge of concentrated bleach into a wastewater stream was eliminated by installing a continuous flow, closed-loop tank system. Bleach continuously flows through the loop and is sent to the reaction vessel (where it is completely reacted) only when needed.	Bleach (N)	100%	10,000 gal/yr (83,160 lb/yr)
Solid (OP) 1983	By replacing pump seals, the amount of soil contaminated with toluenediamine (TDA) due to leaks has been reduced.	TDA-contaminated soil (H)	59% from 1981 to 1987	2,350 lb/yr
Solid/water (OP) 1982	By working with suppliers to provide higher-grade raw materials, significant yield increases and cost reductions have been realized.	Varied (N)		
Water (PS, CH) 1990	The process chemistry has been changed for three products so that nonregulated chemicals can be substituted for regulated ones.	Varied (H)	100%	

Key to source reduction types: CH, chemical substitution; EQ, equipment change; OP, operational change; PR, product change; PS, process change.
A blank indicates that the plant did not provide information.

research. In 1990, after a year and a half of research, further changes in the process chemistry eliminated the need for TCB.

Another source reduction effort at the plant, in 1985, eliminated a concentrated bleach discharge into a wastewater stream. A continuous flow, closed-loop tank system was installed, in which the bleach continuously flows through the loop and is sent to the reaction vessel (where it is completely reacted) only when needed. Previously, a plug of concentrated bleach entered the wastestream whenever the bleach delivery system was turned on to begin a new batch reaction. The change was made because of concern for the safety of workers who might have to come into contact with the concentrated bleach in the wastestream. Bleach is inexpensive, so the dollars saved by eliminating this wasted raw material were considered insignificant.

PMC collects and disposes of soil contaminated with toluenediamine (TDA) from pump leaks. By replacing pump seals at a cost of $10,000, the company reduced TDA-contaminated wastes from 4,000 pounds in 1981 to 1,650 pounds in 1987 (a 59 percent reduction). In part

Change in Yield	Dollars Saved	Dollars Spent	Motivation	Comments	Time Needed for Implementation
		$25,000	Pressure from overseas competitors forces PMC to look very carefully at losses due to waste generation.	Modifications were needed to prevent the generation of mixed solvents as well as to recover individual solvents.	6 mo
			Stricter limits on discharge to municipal sewage treatment plant.	This change resulted in total elimination of TCB at this facility.	1.5 yr
			The change was made because of concern for the safety of workers who might come into contact with the concentrated bleach in the wastestream.	Bleach is cheap so cost savings are insignificant.	
	$10,000/yr		Safety concerns; and need to eliminate hazardous waste.	No cost savings because staff time had to be spent to find non-leaking pump seals for hot liquid.	
+10%	$250,000/yr		In the early 1980s, some yields had declined as much as 5% due to both process problems and raw material impurities.	Raw material purities have increased from 98 to 99.5%.	1 yr
			Federal pretreatment regulations for organic chemical plants.	One product may have to be dropped to meet new requirements.	

because the company had difficulty finding a nonleaking pump for hot liquids, time and staffing costs incurred for the research and installation of the seals outweighed the reduced waste disposal costs. However, safety concerns and the need to eliminate hazardous wastes led to the installation of the pump seals in 1983.

Since 1982, PMC has realized significant yield increases and cost reductions by working with its suppliers to provide higher-grade raw materials. Purity has been increased from 98 to 99.5 percent for many of its raw materials, and waste has been reduced as a result of yield improvement. In addition, cost savings have resulted from reduced purchases of materials. In the early 1980s, some yields had declined as much as 5 percent due both to process problems and raw material impurities. Between 1982 and 1987, yields increased by as much as 10 percent and cost savings were $250,000.

Motivated by regulations for organic chemical plants scheduled to take effect by the end of 1990, the plant eliminated wastes of chemicals regulated under the federal Clean Water Act. The process chemistry has been changed for three products so that nonregulated chemicals can be substituted for regulated ones. The plant expected to drop one product in order to meet the new pretreatment regulations.

Other Waste Management Practices

PMC reported several waste management practices in addition to its source reduction activities. Before the plant eliminated TCB waste altogether in 1990, TCB waste used to be finely dispersed in roughly 15 to 20 separate wastestreams. A $250,000 investment brought all of these wastestreams to a common point, where the chemical was coalesced, decanted, and returned to the process. Recovered TCB reduced raw material costs by roughly $100,000 per year, representing a reduction of 150,000 pounds per year.

In anticipation of regulations imposed by the municipal treatment plant to which PMC's wastewater is discharged, PMC reduced the copper concentration in its wastewater from a range of 15-35 ppm to less than 10 ppm, and frequently less than 2 ppm. The municipal authority, in an effort to satisfy Ohio EPA, debated this matter for 3 to 4 years, giving PMC time to evaluate its own situation and do the research to solve the problem. PMC installed a copper recovery system at a cost of $300,000. Savings are not yet known because PMC has not determined just what can be done with the recovered copper. The company is looking into reuse options as well as landfilling or selling the impure copper back to the supplier. Various techniques for doing this have been demonstrated in the laboratory but not yet on an operational basis at the plant.

This plant is located in an air nonattainment area; that is, ambient air standards set by the Clean Air Act are not being met for the area in general. The state did a RACT (reasonable available control technology) study for PMC's air permits in 1990. PMC reported that, while the improved efficiency of their processes has reduced air emissions, they will probably add scrubbers for ammonia recovery to meet the new permit standards.

Technical Assistance

In the early 1980s, Sherwin-Williams' Environmental and Safety Group served the corporate divisions. With only five staff members, and with the Cincinnati plant supplying only 5 percent or less of the company's sales, not much attention was given to this plant. Therefore, the Cincinnati plant developed its own on-site technical expertise. Since that time, according to the plant manager, the plant has developed a strong technical team with people who know the system well. Information is also shared between PMC's three locations.

The plant manager also finds technical information exchange through such trade associations as the Ohio Chemical Council, the Ohio Association of Manufacturers, and the Chamber of Commerce useful. In addition, raw material suppliers are a source of information. For example, PMC's TDA supplier provided some information on sealing the leaking pumps. At the same time, other companies (for example, Emery and Proctor & Gamble) have asked PMC to provide them with information on using pumps with hot materials.

Company Comments on State and Federal Regulations

PMC reported that, while economic incentives are its major encouragement for source reduction, government regulations force the issue. Company officials noted that the company probably would not have as aggressively pursued copper and TCB reduction without the scrutiny of the municipal sewage treatment plant operators.

PMC finds Ohio EPA's policies on air emissions confusing. It has found that interpretations within the Ohio EPA can differ. PMC had not heard of Ohio EPA's grant program for source reduction projects. However, it did express interest in that type of program.

In general, PMC believes that as long as the company is in compliance with regulations, there is no reason to contact a government agency. However, without a full-time person reviewing federal and state actions, it finds it hard to keep up with new and changing regulations.

RHÔNE-POULENC, INC.
New Brunswick, New Jersey

Summary

Rhône-Poulenc's New Brunswick facility is a relatively old facility, built in 1949, with mature processes and equipment, located 25 miles southwest of Newark, New Jersey. Rhône-Poulenc, S.A. is France's largest nationalized chemical company, and the New Brunswick facility is one of eleven US subsidiaries. Production decreased at the facility during the 1980s. In 1982, the production of bulk pharmaceuticals was discontinued, and in 1985 production of two aroma chemicals and their intermediates was also discontinued. Furthermore, the plant has discontinued its fragrances and flavors line. The plant now produces only aroma chemicals, rare earth compounds, and chemical intermediates such as salicylaldehyde which is used in making coumarin and metal deactivators. The facility size has been reduced through the sale of 5 of its 20 acres, and a warehouse that handled imports has been relocated elsewhere. The number of employees has been reduced by 40 percent from just less than 100 in 1981 to just over 60 workers in 1989.

A reorganization of the parent company led to an environmental policy emphasizing the development of clean technologies for use in all its plants. This approach stems from the regulatory structure of France where worker, environmental, and public health safety are not separately categorized as they are in the United States. The corporate policy also requires each of its plants to have a "waste minimization program" that seeks to recover valuable materials and reduce the volume of waste generated. Because this Rhône-Poulenc plant has older, mature processes dating to the 1960s, its source reduction efforts concentrate on optimizing process conditions and reusing raw materials in the processes.

Rhône-Poulenc granted an on-site interview to INFORM and conducted a tour of its New Brunswick facility. The plant reported a total of seven source reduction activities reducing 248,010 pounds of waste and saving the company $365,500 each year. The company did not grant an interview for INFORM's 1985 study.

Products and Operations

Most of the New Brunswick plant's processes date to the early 1960s and 1970s. Currently, the plant produces seven products and two intermediates. The two main products are the aroma chemicals coumarin (sold as a fixative for other fragrances) and ethyl vanillin (synthetic vanilla). The plant also produces salicylaldehyde, which is a chemical intermediate used in making coumarin and metal deactivators.

Production of bulk pharmaceuticals and of two aroma chemicals and their intermediates was discontinued in the 1980s. The primary reason was economics, but the potential for accidents in handling one particular raw material (titanium tetrachloride) was considered in eliminating its use at the plant. Due to the nature of the discontinued operations, wastewater and air emissions have been reduced more than solid wastes.

The facility size has been reduced through the sale of 5 of its 20 acres. A warehouse that handled imports has been sold and this function has been relocated elsewhere. The number of employees has been reduced by 40 percent from just less than 100 in 1981 to just over 60 in 1989.

Environmental Policy

Rhône-Poulenc, an international company, is among the ten largest chemical companies in the world. The company has a written health, safety, and environmental policy. These policies set standards in the areas of environmental matters, personnel safety, and other general standards. Included in the environmental section are formal standards for an environmental program, permit compliance, "waste minimization" (including source reduction and recycling), waste disposal, resource conservation, and release reporting. Each company site is required to have an effective "waste minimization program" to recover valuable materials and to reduce the volume of waste generated. Rhône-Poulenc managers said that the hierarchy that they follow in dealing with hazardous wastes is to first reduce the generation of waste to the maximum extent possible, then to maximize recycling of wastestreams and use of off-site recyclers, and then to utilize safe disposal, with land disposal only as a last resort.

In addition to the company's Waste Minimization Standard, developed in 1988, Rhône-Poulenc is a participant in the Chemical Manufacturers Association's Responsible Care Program and is in the process of incorporating elements of that program's "waste and release reduction code of management practices." These practices are designed to achieve ongoing reductions in the amount of all contaminants and pollutants released to the air, water, and land. An environmental index is also being developed to monitor progress in source reduction and recycling.

The plant itself has a written "waste minimization plan" covering all types of wastes. The 1989/1990 plan states that, in the area and research and development, the plant will:

- determine the optimum reaction and processing conditions to maximize yield and minimize the generation of undesired by-products, and
- develop the technology to optimize the recovery of starting materials for recycle back into the process.

For the production area, the plan requires:

- strict adherence to standard operating procedures to maximize yield, minimize by-product generation, and prevent the production of off-specification products that might require disposal as hazardous waste, and
- maximization of the use of wastestreams as fuel blend rather than disposal as solid waste.

For the laboratory, the plan requires:

- ordering of minimum amounts of laboratory chemicals to minimize later disposal requirements, and
- maximizing the use of recycling of raw materials or products back into the production unit.

The plan also includes recycling for aluminum, paper, glass, and metal drums.

Materials Data Collection

New Brunswick plant management is responsible for complying with and meeting the reporting requirements of local, state, and federal regulations. The corporation also expects the plant to go beyond government regulations and apply process improvements and good operating practices. Plant managers are assisted by the corporate and divisional Health, Safety and Environmental Department, which both monitors plant practices and aids plants in regulatory compliance and facility upgrading. This assistance includes periodic reviews by the corporate staff. The reviews cover compliance with governmental regulations and corporate guidelines and policies, the effectiveness of spill release and reporting requirements, and opportunities for source reduction.

Task forces are used on a periodic basis to perform process reviews. During the review, the process wastes of all types are quantified and characterized with a view to discovering opportunities for improving yields, source reduction, and recycling. They compare disposal records with those of similar processes at other Rhône-Poulenc plants, for example. Task force representatives include plant personnel as well as personnel from the corporate Engineering Department and Process Development Department. The surveys cover individual chemicals in the wastes, as well as waste categories, and include nonprocess operations such as loading/unloading and other material transfers.

This type of review is done every 4 to 5 years at the Rhône-Poulenc New Brunswick plant, with the most recent one in 1991. The environmental review can take up to 9 months. It is Rhône-Poulenc's experience that a multidisciplinary team that includes personnel from outside the plant is essential to the success of such review.

Monitoring of each process to measure yield and to track consumption of raw materials and waste per pound of product is done at this plant. Rhône-Poulenc reported that it had always monitored raw material consumption and, in 1979, instituted waste monitoring. Each raw material has been assigned a usage factor, established by the on-site laboratory and refined by experience, that is used to identify operational control problems.

All costs of waste treatment and disposal are allocated back to processes. These costs include treatment, expenses for regulatory compliance, insurance, spill clean-up, and public/customer relations when dealing with waste issues.

Other Source Reduction Program Features

In addition to a formal source reduction policy, materials accounting (but not materials balance), and cost accounting, Rhône-Poulenc had three of the other program features tracked by INFORM: an environmental program (but not environmental goals), leadership, and employee involvement.

Several different aspects of required corporate management procedures make up the approach to source reduction at this Rhône-Poulenc plant. The plant receives written management directives, for example, on release and spill prevention, regulatory compliance, and disposal site selection. A required report from each plant on implementation of the directives acts as a tool to point out potential problem areas. Rhône-Poulenc's New Brunswick personnel also learn from other plants' efforts through a review of these reports. Since 1989, every Rhône-Poulenc plant is required to complete an annual environmental report that contains a section on source reduction and recycling.

In addition, any major new processes are subject to review by an international corporate group. The review covers the extent to which the proposed process is utilizing "clean technologies" and waste is avoided.

Progress made in source reduction and recycling is the responsibility of the New Brunswick health, safety, and environmental assessment supervisor. A committee has been formed at the plant to look for ways to reduce odor and air emissions of volatile organic compounds (VOCs). Another committee reviewed plans for a pretreatment system for wastewaters that has been approved and designed and is now under construction. Personnel on these committees include operations management, safety, environmental, technical, and engineering employees. The plant management is also assisted by various corporate departments that conduct periodic surveys that include source reduction and characterization of process waste.

Certain safety and environmental research is done in France. There is both a Process Research Department and an Environment Department. In the Process Research Department, the goal in developing new technologies is to find "clean technologies" that incorporate both maximum recycling and equipment designed to minimize losses. This research effort is aided by the work of the Environment Department, whose purpose is to study ways of preventing and controlling the harmful effects of pollution caused by the manufacture and use of chemical products. A staff of 30 people conduct research looking at environmental impact and toxicity studies as well as waste treatment methods. They study both product waste generation and product reaction and transformation in the environment in order to develop products that generate less waste.

The research that goes into the development of a new product includes a look at its impact on the environment and its compliance with environmental protection standards. Both product manufacture and the predicted circumstances of its use are examined. All possible discharges of waste are analyzed, as well as estimated waste distribution between air, water, soil, sediments, and living organisms.

This multimedia focus is a result of the French corporate ownership. Under the regulatory

Table II-33 Rhône-Poulenc (New Brunswick, NJ): Source Reduction Activities

Waste Medium (SR Type) Year	Source Reduction Activity	Specific Waste Reduced (Hazardous or Nonhazardous)	Percent Waste Reduced	Amount Waste Reduced
Water (OP) 1987	All wastewater streams containing toluene are collected into one settling tank. The recovered toluene is put back into the processes through a closed-loop pipe system.	Toluene (H)	40%	100,000 lb/yr
Water (OP, EQ) 1990	Toluene will be collected in tank adjacent to operations unit, recovered through distillation, and reused in salicylaldehyde process.	Toluene (H)	20%	30,000 to 40,000 lb/yr
Air (EQ) 1987	In-line condensers, installed on the salicylaldehyde process, cool air lost during drying and recover the product from the emissions.	Salicylaldehyde (N)		10 lb/batch (average)
Solid/air (EQ) 1982 or 1983	Residues from the ethyl vanillin process were reduced by removing a piece of equipment that was degrading the product.	Ethyl vanillin by-products (N)		50,000 lb/yr
Solid/air (PS) 1982 or 1983	Reduction in volatile organic solvent (VOS) emissions achieved by eliminating use of a VOS.	VOS (N)	100%	50,000 lb/yr
Solid (OP) 1970s	As standard practice, all quality control and raw material samples are sent back to be reused in the production processes.	Various (H, N)		3,000 lb/yr
Water (PS) 1987	Optimization of the salicylaldehyde process has resulted in yield improvements.	Salicylaldehyde and by-products (N)		60,000 lb/yr

Key to source reduction types: CH, chemical substitution; EQ, equipment change; OP, operational change; PR, product change; PS, process change.
A blank indicates that the plant did not provide information.

structure of France, safety means the full spectrum of worker safety, environmental safety, and public health safety. These are not separately categorized as in the United States. Rhône-Poulenc plants in the United States have adopted this approach because it ultimately saves them money.

A reorganization of the parent company in 1975 helped to focus the company on source reduction in two ways. One was the concept of "clean technologies" and integrated cradle-to-grave handling of all materials (something the New Brunswick plant managers consider the French to be especially good at). The second was increased cross-communication among plants and communication from headquarters to individual plants as the corporation focused on its international status and exerted stronger control over the individual plants. Previously, the company considered itself to be a holding company of separate plants with different products and there was less effective communication among the plants.

Change in Yield	Dollars Saved	Dollars Spent	Motivation	Comments	Time Needed for Implementation
	$16,000/yr	$40,000	Periodic process reviews, which identify potential for materials loss, noted that a lot of toluene was being discharged when all waste-streams were looked at.	Toluene is used in large quantities (about 3,500 gallons per day) as a solvent at the plant and losses should be small since it is not a reactant.	
	$4,500/yr	$4,000,000	Federal pretreatment regulations for organic chemical plants.		1 yr
+0.5%	$30,000/yr	$10,000	To improve product yields. Product had been lost during the drying stage.	Odor reduction also achieved.	
	See below	See below	"Waste minimization" and cost reduction.		1 mo
	$45,000/yr in combination with above equipment change.	$10,000 in combination with above equipment change.	"Waste minimization" and cost reduction.		1 mo
	$20,000/yr		"Waste minimization" and cost reduction.		
+2%	$250,000/yr	$200,000	"Waste minimization" and cost reduction.		1 yr

Source Reduction Activities

For INFORM's prior study, *Cutting Chemical Wastes*, Rhône-Poulenc reported no source reduction activities. However, for this study, the plant described seven source reduction measures. These are summarized in Table II-33 and described below.

Overall, between 1981 and 1987, production volume was reduced by 19.5 percent in quantity of finished product. Over the same time period, aqueous wastes were reduced by 35 percent, air emissions by 86 percent, and solid wastes by 45 percent. These figures do not include the one-time clean-up of discontinued processes and the warehouse property sale in 1985. Beyond regulatory requirements, economics has been a prime mover behind the source reduction efforts at this plant, with liability concerns also a major consideration.

The periodic reviews, mentioned above, search for ways to reduce material losses. One example of the result of such a review is the reduction of toluene discharges. Toluene is used in large quantities (about 3,500 gallons per day) as a solvent at the plant, and losses were expected to be small since it is not a reactant. However, the 1983 review noted that a lot of toluene was being discharged taking all the wastestreams together, even though individual wastestreams did not necessarily contain excessive toluene. Since then, at a cost of $40,000, losses of this chemical have been reduced by 100,000 pounds per year for an annual savings of $16,000. This has been accomplished by collecting all sources of toluene and water into one settling tank, and then reintroducing the toluene into the processes. The collection and redistribution of toluene is accomplished through a closed pipe system. For batch processing operations, settling at each source would be time-consuming and expensive and controlling 14 individual settling tanks is operationally more difficult; thus, a separate collector tank is used. This process change has also reduced wastewater treatment requirements.

In another source reduction activity affecting toluene, the New Brunswick plant undertook a $4 million capital project to reduce toluene in wastewater that is discharged to the Middlesex sewage treatment plant. This project is designed to comply with the federal pretreatment regulations for organic chemical plants. The toluene, expected to amount to 30,000 to 40,000 pounds per year, will be collected in a storage tank adjacent to the operations unit, recovered through distillation, and reused in the salicylaldehyde process. This will represent an additional 20 percent reduction in the usage of toluene and a cost savings of $4,500 per year.

In an example of source reduction undertaken to improve product yield, in-line condensers have been installed on the salicylaldehyde process. The operation had been losing product during the drying stage. The condensers cool the air lost during drying and recover the product from the emissions. Reductions in odor were also achieved. The average amount of product recovered is 10 pounds per batch, and yield has increased by 0.5 percent. The installation of the condensers in 1987 cost $10,000 and saved $30,000 in the first year of operation.

By removing a piece of equipment that had been degrading a product, an equipment change in 1982 or 1983 succeeded in reducing solid wastes produced by the ethyl vanillin process by 50,000 pounds per year. At the same time, the use of a volatile organic solvent was discontinued and, therefore, air emissions of the solvent were eliminated. Together, these two changes cost $10,000, took 1 month of research and development, and save Rhône-Poulenc $45,000 per year. They were undertaken to reduce wastes and production costs.

As standard practice at the plant since the 1970s, all quality control and raw material samples are sent back to be reused in production. This practice reduces hazardous and nonhazardous solid waste by 3,000 pounds per year and results in cost savings of $20,000 per year.

In 1987, after a year of research and development, the salicylaldehyde process was optimized, resulting in a 2 percent improvement in yield, $250,000 in annual savings, and 60,000 pounds less waste per year. The cost of this cost and waste reduction effort was $200,000.

An example of an effort, so far unsuccessful, to reduce waste through the application of new technologies to existing processes is Rhône-Poulenc's investigation of technologies to reduce losses of phenol. Several task forces over the years have studied the plant's losses of phenol. However, available reduction technologies are not economically practical because both phenol and phenol treatment are inexpensive (since phenol will not volatilize and is biodegradable). The search is continuing because any loss of a raw material is a source of concern to plant management.

Other Waste Management Practices

Aside from its source reduction activities, Rhône-Poulenc reported a significant reduction in land disposal of wastes that has been achieved, primarily through the use of off-site incinerators or the use of solid wastes off-site as alternative fuels. These changes have been adopted because of the liability associated with land disposal. In 1981, 100 percent of solid wastes from this plant were landfilled; in 1986, 46 percent were landfilled and 54 percent were incinerated; and in 1989, 100 percent of routine solid wastes were incinerated.

Technical Assistance

Technical information is received from within the corporation through, for example, seminar meetings conducted by corporate management. The New Brunswick plant manager reported that he has found the technical information from the seminars useful and that meeting the managers of other plants is also an effective way to learn about new source reduction techniques. While the primary source of information is the corporation, Rhône-Poulenc also uses outside consultants, such as industrial hygienists, from time to time.

Information from state or other government sources is generally not useful to the plant manager because it is not specific enough for this plant's operations. According to plant officials, the easy approaches to source reduction, the usual focus of government efforts, have already been taken. Additionally, they have found it hard to find out who in government has the information they need.

Rhône-Poulenc has found sharing information outside the corporation difficult because of both the proprietary nature of the data and antitrust laws. On the other hand, Rhône-Poulenc has found that trade organizations do a good job of providing information; the technical sessions of the Chemical Manufacturers Association, for example, have been helpful. The trade groups have rigid rules regarding proprietary information and their programs are useful in learning from other companies. The New Brunswick plant manager believes that trade groups should sponsor more technical seminars.

Company Comments on State and Federal Regulations

In addition to regulatory requirements, plant officials reported that economics are a primary factor in the source reduction efforts at the Rhône-Poulenc New Brunswick plant. However, they prefer tax credits or loan and grant programs to mandatory legislated reductions in waste generation. They note that the problem with mandatory reductions is that it is much easier to comply if a plant has not done much in the way of source reduction, but it is hard if steps have already been taken.

Future

Plant officials expect small gains at the New Brunswick plant because the current process chemistry is set, although Rhône-Poulenc continues to look at new processes for possible application to this plant. The primary focus in the next few years will be on reducing odor-related and VOS (volatile organic substances) emissions.

SCHER CHEMICALS, INC.
Clifton, New Jersey

Summary

The Scher Chemicals facility in Clifton, New Jersey, just north of Newark, is the company's only plant. The company started in the 1930s as a chemical distributor, but did not start manufacturing chemicals until this plant was built in 1956. It manufactures a wide variety of speciality chemicals for use in cleaning products and in the cosmetics and textiles industries. Scher manufactures more than 200 products and has no plans to expand into other areas. Part of a matured industry, it has 20 employees at the plant. During the 1980s, the amounts of chemicals used and produced changed as a result of changes in demand: production of the surfactant epichlorohydrin doubled, while demand for acrylonitrile dropped.

Scher's general policy is to reuse or recycle any by-products and off-quality batches. Scher discontinued use of several hazardous chemical intermediates and let its customers know it would no longer supply the products since volume was small and the risks of using the chemicals in a residential neighborhood were not worth continued use.

Scher granted an on-site interview to INFORM and conducted a tour of its Clifton, New Jersey, facility. The plant reported one source reduction activity. The company had granted an interview for INFORM's 1985 study, but reported no source reduction activities at that time.

Products and Operations

The Clifton plant manufactures more than 200 products including surfactants, emollients, water softeners, detergents, lubricants, and speciality coatings for use in the manufacture of glass bottles. It has a sufficient variety of products to meet customer demands and no plans to expand into other areas. Research is focused on how to increase yield and decrease production time, and on testing the chemicals of potential substitute suppliers of the raw materials.

The amounts of the chemicals used and produced has changed since the early 1980s. Demand for acrylonitrile has dropped off so that less overall waste is produced at the plant now compared to the early 1980s. The demand for maleic anhydride is about the same. Scher no longer uses tetrachloroethylene, which had been used to perform extraction analysis to evaluate the washing efficiency of a customer's cleaning processes. Scher no longer provides this service and has no other use for tetrachloroethylene.

Since the early 1980s, Scher has increased production capacity at the plant with the purchase of new equipment and increased storage facilities. The major product increase has been in ampiferric surfactants, which use epichlorohydrin. About twice as much is now produced.

Environmental Policy

There is no official environmental policy for this plant. The plant manager oversees all environmental aspects of the operations. Before he was hired, the president/owner of the company had that responsibility. The plant manager's previous experience includes work in the chemical industry for 6 years as a process and project engineer.

Materials Data Collection

The plant manager reports that there are no source reduction inventory procedures at this plant because the hazardous wastes produced are minimal and probably fall into the conditionally exempt category of RCRA (less than 220 pounds per month total hazardous wastes). However, there is an official inventory procedure to track raw materials and products so that they will be used before their shelf life expires. There is no cost accounting program for wastes at the plant.

Other Source Reduction Program Features

In addition to implementing some materials accounting (but not a written source reduction policy, materials balance, or cost accounting), Scher reported fully or partially implementing two other features tracked by INFORM: leadership and an environmental program (but not employee involvement or specific environmental goals).

Source Reduction Program and Activities

The plant manager reports that because the amounts of regulated hazardous waste generated at the plant are so small, it is not feasible to spend time looking for ways to reduce these wastes. Additionally, any by-products are reused or recycled, and off-quality batches are reworked or blended back into products. More than 95 percent of these materials is reported to be reused or recycled. The plant manager has had discussions about possibly selling out-of-date inventory in response to advertisements in trade magazines, but he does not know yet if the small amounts would justify the effort.

The plant's one reported source reduction activity (summarized in Table II-34), took place in 1989. Scher eliminated the use of two or three hazardous chemicals that had been used as intermediates to make nonhazardous products. It let its customers know that it would no longer be making the products. Some were small volume products, but even for those that were not, Scher decided that the risks of using the chemicals in the residential neighborhood where it is located outweighed the benefits. Scher now only uses two hazardous chemicals: epichlorohydrin and diethylphosphate.

Technical Assistance

The plant manager generally obtains information on changes in government regulations that might affect his operations from other companies. Dow Chemical, the supplier of epichlorohydrin, provided assistance on how to reduce reactor emissions. Most large producers publish booklets that contain guidelines on how to control emissions or treat wastes produced when using their products, and Scher was able to use Dow's discussion of epichlorohydrin to make a caustic scrubber on a vent line in the plant. The resulting scrubber solution is a nonhazardous mixture of salts and water. Also, when epichlorohydrin is in use, Scher monitors the plant to make sure the ambient levels are below OSHA standards.

The plant manager said that, while large companies have the expertise to know how to reduce or control chemical wastes, assistance to small companies from the government would be welcome. He thought that a government bulletin on how the average small company can reduce waste, or a bulletin with guidelines on possible approaches, would be helpful. However, he emphasized that it would only be helpful if there was a single list of guidelines because so often the state and federal governments provide conflicting or confusing information and are not coordinated in their approach to helping a plant.

Scher's plant manager would much rather see information on how to comply with government regulations come from the government. He believes that this would be better than the current situation in which a private concern interprets what the government wants and then uses its own interpretation to tell the plant how it can meet the government's standards.

Company Comments on State and Federal Regulations

The plant manager is aware of the state regulations requiring reporting of wastes and production levels. He is not aware of any information the state might have on source reduction. There have been no fees assessed at the plant other than the sewage treatment plant user fees.

Table II-34 Scher Chemicals (Clifton, NJ): Source Reduction Activities

Waste Medium (SR Type) Year	Source Reduction Activity	Specific Waste Reduced (Hazardous or Nonhazardous)	Percent Waste Reduced	Amount Waste Reduced
(PR) 1989	Eliminated two or three products that used hazardous chemicals as intermediates.	(H)	100%	

Key to source reduction types: CH, chemical substitution; EQ, equipment change; OP, operational change; PR, product change; PS, process change.
A blank indicates that the plant did not provide information.

Change in Yield	Dollars Saved	Dollars Spent	Motivation	Comments	Time Needed for Implementation
			Use of hazardous chemicals in their residential neighborhood not worth the risks.	Products were non-hazardous but chemicals used to make them were hazardous.	

SHELL CHEMICAL COMPANY
Martinez Complex, Chemical East
(Criterion Catalyst)
West Pittsburg, California

Summary

The Shell Chemical Company operations in West Pittsburg, California, built in 1931, are part of Shell Oil Company's Martinez Complex, which includes a petroleum refinery. The West Pittsburg plant is one of Shell Oil Company's nine US chemical manufacturing facilities. It manufactures inorganic metallic catalysts for dehydrogenation and hydrotreating processes. In 1983, the plant had 59 employees.

Shell reported under federal and state requirements that it has a written policy on source reduction and recycling, but while source reduction opportunities were identified during 1987, they were not implemented. Also, Shell reports to the federal government show that RCRA waste in particular, and also off-site transfers and surface water discharges, increased from 1987 to 1988.

Shell declined to be interviewed for this study and did not provide any written information. Data on releases and transfers of waste and the plant's environmental program are from reports filed by Shell with the state of California and the US Environmental Protection Agency.

Shell did not grant an on-site interview to INFORM for this study or for INFORM's 1985 study. There is no indication that source reduction measures have been taken at the West Pittsburg, California plant.

Products and Operations

The Shell Chemical Company operates a small chemical manufacturing facility as part of the Martinez manufacturing complex, the site of Shell's Martinez oil refinery. Shell Oil is owned by Royal Dutch/Shell Group, a holding company owned by Royal Dutch Petroleum Company (60 percent), a Netherlands company, and Shell Transport and Trading Co., Ltd. (40 percent), a British company. Shell Oil's Chemical Division is the seventh largest US chemical company.

Environmental Policy

According to Shell's 1987 RCRA Waste Minimization Report to EPA, the plant has a "written policy or statement outlining the goals, objectives and methods for source reduction and recycling." No further information is available.

Materials Data Collection

The 1987 RCRA report also states that site-wide source reduction and recycling inventories were first conducted in 1986. Process-specific inventories to identify opportunities to reduce the generation of hazardous wastes had not been conducted as of 1987 at this site.

Because Shell did not supply INFORM with data on its waste generation or reduction, data from governmental sources are presented. Table II-35 shows the releases and off-site transfers of chemicals as reported to the Toxics Release Inventory (TRI) in accordance with Section 313 of the Superfund Amendments and Reauthorization Act. This information is required for certain chemicals used or manufactured at an industrial facility. In 1987, Shell reported on 21 chemicals, with releases recorded for 18. In 1988, the use or manufacture of 20 chemicals was reported (lead was no longer reported), with releases recorded for 17. From 1987 to 1988 there was a 1 percent increase overall in releases and transfers of TRI chemicals at this facility. While air emissions decreased 36 percent, off-site transfers increased by 43 percent and discharges to surface waters increased by 49 percent.

Table II-35 Shell Chemical (West Pittsburg, CA): TRI Toxic Chemical Releases and Transfers, 1987 and 1988

Chemical	Air Emissions (lb/yr)	Surface Water Discharges (lb/yr)	Off-Site Transfers (lb/yr)	Total (lb/yr)
1987				
Aluminum oxide	140,250	0	153,000	293,250
Ammonia	500	7,600	0	8,100
Asbestos (friable)	0	0	100,000	100,000
Benzene	13,900	0	0	13,900
Chlorine	250	0	0	250
Cyclohexane	5,650	0	0	5,650
Diethanolamine	0	0	0	0
Ethylbenzene	5,200	0	0	5,200
Ethylene	12,250	0	0	12,250
Lead	250	250	0	500
Methyl ethyl ketone	8,574	0	0	8,574
Naphthalene	7,950	0	0	7,950
Phosphoric acid	0	0	0	0
Propylene	30,250	0	0	30,250
Silver	0	64	0	64
Sodium hydroxide	0	0	0	0
Sulfuric acid	250	0	0	250
tert-Butyl alcohol	1,000	0	0	1,000
Toluene	44,000	0	0	44,000
1,2,4-Trimethylbenzene	3,400	0	0	3,400
Xylene	14,500	0	0	14,500
Total	288,174	7,914	253,000	549,088
1988				
Aluminum oxide	34,640	0	309,512	344,152
Ammonia	12,450	10,200	0	22,650
Asbestos (friable)	0	0	51,790	51,790
Benzene	6,400	250	0	6,650
Chlorine	1,000	0	0	1,000
Cyclohexane	2,528	0	0	2,528
Diethanolamine	0	0	0	0
Ethylbenzene	8,713	250	0	8,963
Ethylene	43,748	0	0	43,748
Methyl ethyl ketone	8,574	0	0	8,574
Naphthalene	3,720	0	0	3,720
Phosphoric acid	0	0	0	0
Propylene	13,780	0	0	13,780
Silver	2,084	250	250	2,584
Sodium hydroxide	0	0	0	0
Sulfuric acid	250	0	0	250
tert-Butyl alcohol	1,000	0	0	1,000
Toluene	15,100	250	0	15,350
1,2,4-Trimethylbenzene	5,689	250	0	5,939
Xylene	23,200	250	0	23,450
Total	182,876	11,700	361,552	556,128

Source: SARA Title III, Section 313, Toxics Release Inventory (TRI) Report Form R for 1987 and 1988.

Table II-36 shows that RCRA wastes alone virtually doubled in both 1986 and 1987. Shell reported that it requested to have its Part A permit application (to recycle, treat, store, or dispose of hazardous wastes on-site) withdrawn in 1987.

Table II-36 Shell Chemical (West Pittsburg, CA): Generation of RCRA Hazardous Wastes, 1985-1987

Year	RCRA Waste (lb)	Change from 1985 (%)
1985	1,424,798	
1986	4,143,880	+191%
1987	9,246,000	+549%

Sources: 1985, average of 1985 values reported in the Facility Biennial Report and its Hazardous Waste Information System 1985 Summary; 1986, Hazardous Waste Information System 1986 Summary; 1987, Facility Hazardous Waste Report for 1987.

Other Source Reduction Program Features

Shell's 1987 RCRA Waste Minimization Report shows that the West Pittsburg chemical plant established a source reduction and recycling program prior to 1986, and expanded the program in both 1986 and 1987. Capital expenditures for source reduction and recycling were $50,000 prior to 1986 and $400,000 during 1986.

Shell Oil's 1987 annual report states that "earnings of our Chemical Products segment were the best in its history." However, no capital expenditures were made for source reduction and recycling in 1987. Furthermore, the Chemical Products segment reportedly generated $433 million in surplus cash above capital expenditures, yet no operating costs devoted to source reduction and recycling of hazardous wastes were reported in either 1986 or 1987.

A "Quality Improvement Process" was instituted at all of Shell Oil's product organizations in 1984. The program "stresses a philosophy of doing each job right the first time, encouraging employees to find ways to perform their work better and eliminate waste and inefficiency." The program does not provide specific incentives for employees to identify and implement source reduction and recycling opportunities and activities. Shell's 1987 annual report states that the program has "resulted in significant cost reductions by improving productivity."

Source Reduction Activities

The 1987 annual report goes on to say that source reduction opportunities to reduce the volume and/or toxicity of hazardous waste generated were identified in 1987, but were not implemented. No source reduction opportunities were identified in 1986 or in prior years. In 1987, opportunities were identified to recycle hazardous wastes but, again, were not implemented.

Part II of Shell's 1987 RCRA Waste Minimization Report states that no source reduction and recycling results were achieved in 1987. None of the Toxics Release Inventory report forms submitted by this Shell facility in 1988 contained any source reduction information in the optional section provided for this purpose.

Technical Assistance

According to Shell Chemical East, the site requested or received technical information or financial assistance on source reduction and/or recycling practices from state and federal government, trade associations, other parts of Shell Oil, and other sources, including conferences and literature.

UNOCAL CHEMICALS
(formerly Union Chemicals Division)
La Mirada, California

Summary

The Unocal Chemicals plant in La Mirada, California, east of Los Angeles, was built in 1949, and has two operating units: a latex polymer production unit and a solvent blending operation. The latex polymers are used in carpet backing, concrete adhesives, and paint formulations. The solvent blending operation takes chemicals stored in underground tanks and mixes them in batches according to customer specifications. Solvent production increased about 20 percent from 1985 to 1988. However, because of benzene's recognized carcinogenicity, Unocal discontinued use of this solvent at the La Mirada plant, consolidating its use at a single regional facility.

Unocal's corporate environmental policy stresses compliance with environmental regulations but does not mention source reduction. All waste generated from the solvent blending operations takes the form of air emissions. These are difficult to measure directly, and Unocal uses a formula based on production amount and the type of storage tank, among other factors. Because these measures are indirect and because Unocal is meeting all applicable air regulations, the company has little incentive to pursue source reduction in connection with these operations.

INFORM was able to obtain an on-site interview for the 1985 report. However, while the plant manager of the solvent blending operations granted INFORM an on-site interview for this report, the plant manager of the latex operations did not. Unocal reported no source reduction activities at this time, although it had described one for the earlier report.

Products and Operations

Unocal is a large firm best known for marketing petroleum products under the label "Union 76." The La Mirada plant is one of six petrochemical plants owned and operated by Unocal. There are two completely distinct operations at the La Mirada plant: latex polymer production and solvent blending.

Latex polymer production at La Mirada commenced in 1967, with the primary products being polyvinyl acetate and styrene-butadiene latexes (used for carpet backing, concrete adhesives, and paint formulations) and acrylic resins (used in acrylic paint formulations). In 1985, approximately 50 million pounds of polymers were produced annually in water-based batch process operations that operated 24 hours a day, seven days a week. No updated information was available for these operations.

The solvent blending operations at the plant involve a wide variety of organic chemical solvents that are stored in underground tanks and mixed, resulting in a plethora of different blends. In some instances, individual solvents may be drummed and sold to customers unblended. All blending operations are done in batches to customer specifications. Paint companies comprise the majority of the plant's customers, although janitorial supply companies have recently started to seek solvents from Unocal as well. Production levels range from 25 to 30 million gallons per year, an increase of 15 to 20 percent over 3 years. The organic chemicals handled at the plant include: methanol, methyl ethyl ketone, methyl isobutyl ketone, perchloroethylene, toluene, dichloromethane, 1,1,1-trichloroethane, acetone, cyclohexane, hexane, textile solvents, and rubber solvent. The La Mirada plant formerly used benzene but, because of its recognized carcinogenicity, Unocal discontinued use of this solvent at the La Mirada plant and consolidated benzene use at a single facility for each region of the country.

Environmental Policy

Environmental and operational responsibilities are integrated at the La Mirada plant. The corporate manager of environmental affairs serves as a resource to the plant manager and coordinates company-wide efforts with regard to underground chemical tanks.

Unocal's explicit environmental policies stress the need to comply with all applicable pollution control regulations. Company guidelines were issued with respect to the hazard warning mandate associated with California State Proposition 65. There is no specific mention of source reduction in corporate environmental policy statements.

Environmental costs are budgeted facility-wide, based upon an extrapolation from the previous year. They are not allocated back to individual products and processes.

Materials Data Collection

Under a corporate compliance program, a review team inspects each facility periodically to monitor compliance with applicable environmental regulations.

In addition, Unocal has installed a $900,000 computerized materials tracking system. The plant manager reports that this system has resulted in better product quality and reduced labor costs.

Source Reduction Program and Activities

While one source reduction activity was reported by Unocal to INFORM for the 1985 study, none were reported for this study. Instead, several obstacles to source reduction were reported at the Unocal plant.

First, the actual amount of waste generated is unknown. All wastes generated are lost to the air; there are no water or solid wastes associated with the solvent blending operation. Because all emissions are to the air and the mixing operation is outside, the plant finds it very difficult to get direct measures of the amount being wasted. All efforts to measure waste generation at this plant are based upon a formula used by the South Coast Air Quality District to monitor the plant. The plant does not know the relationship between the result of this calculation and reality and is merely following the dictates of the local regulatory body in using the formula.

Second, because of the way the formula is defined, source reduction measures would not always change the results of the calculation. The formula only applies to the storage tanks at the plant and involves the following parameters: vapor pressure of the solvent, molecular weight of the solvent, tank throughput in barrels, number of times the tank is emptied and refilled, and a constant factor based upon the specific design of the tank. As a result, only source reduction activities involving changes to the tank that would modify the constant factor, changes in the amount of solvent passing through the tank, or changes in the type of solvent in the tank would make any change in the calculated result. Any other type of source reduction measure, such as those involving changes in seals or other measures to stop leaks, would result in no change in the calculated figure.

Finally, independent of these calculations, the plant manager knows from the computerized materials tracking system that the plant is getting the solvents it buys into product drums without significant losses. Given this conclusion, the fact that the plant is meeting all applicable air regulations, and the low cost of these materials, the manager believes he has little incentive to pursue the issue of loss reduction any further.

Other Waste Management Practices

Unocal reported two other waste management concerns. First, although calculations using the South Coast Air Quality District formulas result in predicted air emissions that fall below thresholds set by the district for vapor recovery and emissions control measures, if the solvent blending operations continue to grow, the result of the calculations may exceed the applicable thresholds. This would result in the plant having to install vapor recovery and emissions control equipment.

Second, the facility is facing a major decision about what to do with its underground storage tanks that were installed in 1968 and are therefore nearing the end of their useful life. New regulations of underground storage tanks ban their use after 1995. In the interim, the company finds it very difficult to ensure that no underground contamination problem will occur with tanks of this age. The company must decide whether to replace the tanks or to discontinue their use now.

Technical Assistance

The manager of the solvent blending operations expressed the view that the process being used at the plant was straightforward and had been "fine tuned" long ago. Therefore, he expressed no need for outside technical assistance. On issues such as the handling of the underground tanks, he has the corporate environmental manager to support him.

Company Comments on State and Federal Regulations

State or federal policies have virtually no impact on the solvent blending operation, as few pollution control mandates apply to the plant. In fact, the solvent operations do not even have to report to the Toxics Release Inventory as these operations fall outside the manufacturing SIC codes covered by the reporting provision of the Superfund Amendments and Reauthorization Act.

Future

The impending need to address the underground storage tanks at the facility, the solvent blending manager reports, is going to cause the plant to make a careful evaluation of its options in the face of an expanding market for its products.

APPENDIX A

Statistical Tests Performed on Program Feature Data

As part of this research, INFORM examined whether individual study plants had adopted eight specific source reduction program features: a written source reduction policy, cost accounting, materials accounting, materials balance, leadership, employee involvement, specific environmental goals, and an environmental program. These features are discussed in Chapter 2, in the section entitled "Program Features and Plant Characteristics," and Table I-5 on page 28 summarizes source reduction program information for all the study plants.

INFORM performed two statistical tests on the program feature data in order to assess whether the number of source reduction activities reported at a plant was correlated with the adoption of specific source reduction program features and to determine whether two program features were associated with each other or appeared independently.

Correlation of Source Reduction Activities and Program Features

Student's t-distribution was used to test whether the number of source reduction activities at a plant was different, on average, depending on whether the plant had fully adopted the program feature, had partially adopted the program feature, or had not adopted the program feature at all. The test was applied to each program feature individually for those plants reporting on the program feature.

This statistical test examined whether the difference between the mean number of source reduction activities at plants that had adopted an individual program feature and the mean number of source reduction activities at plants that had not adopted the feature was significant at the 95 percent confidence level. That is, it determined whether there was enough variation in the mean number of source reduction activities at the two sets of plants that, 95 times out of 100, such a difference would not show up if random sets of plants were chosen from the universe of all such plants.

To assess the significance of the differences between means, the test calculated the variation in the number of source reduction activities within each set of plants (plants with a particular program feature and plants without it), as well as the mean number of source reduction activities for each set of plants. Because the INFORM sample of plants was small, to perform the test it was necessary to assume that the number of source reduction activities at plants in general is normally distributed.

Table I-6 on page 31 summarizes the results of this statistical analysis: the mean number of source reduction activities at plants with and without each individual program feature and whether or not the differences between the means are statistically significant. As an example, 18 plants reported a full environmental program and 7 reported no environmental program. However, even though the mean number of source reduction activities at plants with no

Table A-1: Correlation between Pairs of Program Features at INFORM Study Plants (Chi-squared test at 95% confidence level)

	Environmental Program	Source Reduction Policy	Cost Accounting	Materials Accounting	Materials Balance	Leadership	Employee Involvement	Environmental Goals	Size	Batch or Continuous
Environmental program	—	I	A	A	I	I	A	A	A	I
Source reduction policy	I	—	A	I	I	I	I	A	I	I
Cost accounting	A	A	—	I	I	A	A	I	I	I
Materials accounting	A	I	I	—	A	I	I	I	I	I
Materials balance	I	I	I	A	—	I	I	A	I	I
Leadership	I	I	A	I	I	—	I	I	I	I
Employee involvement	A	I	A	I	I	I	—	I	I	I
Environmental goals	A	A	I	I	A	I	I	—	I	I
Plant size	A	I	I	I	I	I	I	I	—	I
Batch or continuous process	I	I	I	I	I	I	I	I	I	—

I, independent (two features are not associated on average at a plant)
A, not independent (two features are associated on average at a plant)

environmental program was 4.2, while the mean for plants with full environmental programs was 9.3, the Student's t-test indicated that this difference of 5.1 was not statistically significant at the 95 percent confidence level.

Correlation between Program Features

Many of the program features showed no significant difference, according to Student's t-test, in the number of source reduction activities reported depending on whether the program feature was adopted or not. While in part this was due to the small sample size, it may also have been due to possible correlations between program features. That is, some program features may be prerequisites for others. For example, a materials balance may not be possible without at least some form of materials accounting at the plant. The possibility of correlation or association between each pair of program features was tested using a two-way (or three-way) contingency analysis and the Chi-squared test for independence.

This analysis identified whether the differences between the number of plants with any two program features (for example, both an environmental program and a cost accounting system), the number having neither, and the number having one but not both of these program features (two-way analysis) or the number having partially adopted the program features (three-way analysis) were statistically significant at the 95 percent confidence level. That is, it determined whether, 95 times out of 100, the same variation would not have occurred in a hypothetical set of plants with just normal variation and no association between the two program features.

The Chi-squared test was applied to each pair of program features: environmental program and source reduction policy, environmental program and cost accounting, cost accounting and materials accounting, etc. If the difference between each pair and the hypothetical set of plants was small (as measured against the Chi-squared statistic at the 95 percent confidence level), then the two program features are considered statistically independent. If the difference is large, the two features are not independent and may be associated at the study plants. Table A-1 presents the results of this analysis.

It is important to note that the Chi-squared test only shows whether, on average, two program features tended to occur together at the study plants; it does not show cause and effect. For example, even though environmental program and cost accounting are not independent, according to the test, it cannot be concluded that having an environmental program leads to having a cost accounting system, or vice versa. Instead, the test points to where to look further at the plants and their program features; for example, by asking the companies which program feature they adopted first, and why.

APPENDIX B

Methodology for Estimating Impact of Source Reduction on Waste Generation and TRI Releases and Transfers

In order to estimate the impact of source reduction activities at the study plants on their waste generation and Toxics Release Inventory (TRI) releases and transfers, INFORM made several assumptions and carried out a series of calculations (as discussed in the findings of this report). This appendix presents the background information used for this analysis.

First, for each of the 16 study plants that reported TRI releases and transfers for 1988, Table B-1 shows the amount of waste reduced through source reduction activities reported to INFORM and the amount of TRI releases and transfers reported to the EPA. Only chemical constituents are reported to TRI, while the waste reduction reported to INFORM represented the total amount of waste. Thus, the amount reported to INFORM includes the amount of chemical constituents reduced by the plant, the amount of chemical constituents of waste avoided entirely through process design (that is, never generated as waste), and other wastes reported reduced (RCRA wastes and wastewaters).

Of the other wastes, INFORM selected 10 percent as a reasonable but conservative estimate of constituent chemical waste, or 8.4 million pounds. Adding this to the 3.9 million pounds of constituent chemicals reduced yields a total that can be compared to TRI figures — 12.3 million pounds of constituent chemical wastes reduced through source reduction activities at the 16 plants that reported both to TRI and to INFORM.

Having estimated the amount of waste reduced through source reduction activities that could be compared to TRI figures, INFORM sought to evaluate the impact of the source reduction activities on the amount of TRI chemical waste generated at the study plants. Table B-2 details the steps INFORM used to estimate the amount of TRI waste that would have been generated if there had been no source reduction at the study plants.

According to the EPA's report on the 1988 Toxics Release Inventory,[1] reported TRI figures represented 66 percent of the TRI chemical waste actually generated by the reporting plants since the plants were only required to report the amount of constituent waste released and transferred after on-site treatment. Assuming this relationship holds for the INFORM plants, waste generation of TRI chemicals at reporting plants is actually, on average, 51.5 percent greater than the 24.6 million pounds of reported TRI releases and transfers, or 37.3 million pounds.

---

1 US Environmental Protection Agency, *Toxics in the Community: The 1988 Toxics Release Inventory National Report,* Washington DC, 1990, p. 306.

Table B-1 Waste Reduced through Source Reduction Compared to Reported TRI Releases and Transfers, 1988

Plant	Waste Reduced through Source Reduction			1988 TRI Releases and Transfers (lb)
	Constituent chemicals reduced (lb)	Process design reduction of constituents (lb)	Other reported waste reduced (lb)	
American Cyanamid			805,600	300,125
Aristech			26,929,000	6,168,244
Atlantic			350,000	35,933
Borden			293,070	18,363
Chevron	140,000		0	185,064
Ciba-Geigy			293,000	295,650
Def-Tec			7,535	0
Dow*	1,720,000		0	765,619
Du Pont	0	14,750,000	24,540,000	989,015
Exxon	681,810		16,408,000	591,478
Fisher			629,670	342,949
ICI Americas			125,566	11,134
Merck			12,963,000	10,080,044
Monsanto	1,400,000	15,600,000	329,900	679,913
PMC			90,510	3,050,300
Rhône-Poulenc			248,010	1,098,132
Totals	3,941,810	30,350,000	84,012,861	24,611,963

Constituent waste reduced:	3,941,810 lb
10 percent of other wastes reduced:	8,401,286 lb
Waste reduction comparable to TRI waste:	**12,343,096 lb**

Process design reductions:	30,350,000 lb
Total constituent waste reduced:	42,693,096 lb

* Dow calculation for constituent: (160,000 lb HCl X 0.25)+(12,000,000 lb NaOH X 0.14); therefore, difference in total pounds is made up of water. The pounds of water are not included here because the purpose is to calculate constituent quantities only. (Calculated only for Dow because other plants reported constituent waste.)

Table B-2: Effect of Source Reduction Activity on the Amount of TRI Chemicals Generated at the INFORM Study Plants

1. TRI releases and transfers of INFORM study plants x 1.515 = Amount of constituent waste generated by these plants:

   24,611,963 × 1.515 = 37,290,853 pounds of constituent waste generated by the study plants

2. If no source reduction had taken place, the INFORM study plants would also have generated the 12.3 million pounds of TRI chemicals reduced:

   Pounds of constituent waste at study plants + pounds of TRI chemicals reduced

   37,290,853 + 12,343,096 = 49,633,949 pounds of constituent waste with no source reduction

3. Therefore, the percent of waste that would have been generated had there been no source reduction at the study plants is:

   [(Pounds of constituent waste with no source reduction − pounds of constituent waste generated by the study plants) ÷ pounds of constituent waste generated by the study plants] × 100

   [(49,633,949 − 37,290,853) ÷ (37,290,853)] × 100 = +33.1 percent increase in amount of constituent waste that would have been generated by the study plants had there been no source reduction

4. Also, if no source reduction had taken place, the INFORM study plants would have released or transferred after on-site treatment (if any):

   Pounds of constituent waste with no source reduction × 0.66

   49,633,949 × 0.66 = 32,758,406 pounds of TRI substances that would have been released or transferred had there be no source reduction

5. Therefore, the percent of TRI substances that would have been released and transferred had there been no source reduction at the study plants is:

   [(Pounds of TRI substances if no source reduction − TRI releases and transfers of study plants) ÷ TRI releases and transfers at study plants] × 100

   [(32,758,406 − 24,611,963)/(24,611,963)] × 100 = +33.1 percent increase of TRI substances that would have been released and/or transferred had there been no source reduction

APPENDIX C

Bibliography

Related INFORM publications:

INFORM (David J. Sarokin, Warren R. Muir, Ph.D., Catherine G. Miller, Ph.D., Sebastian R. Sperber). *Cutting Chemical Wastes: What 29 Organic Chemical Plants Are Doing to Reduce Hazardous Wastes.* New York: 1985.

INFORM (Warren R. Muir, Ph.D., Joanna D. Underwood). *Promoting Hazardous Waste Reduction: Six Steps States Can Take.* New York: 1987.

INFORM (Lauren Kenworthy). *A Citizen's Guide to Promoting Toxic Waste Reduction.* New York: 1990. (Revised edition to be published in 1992.)

INFORM (Mark H. Dorfman and John J. Riggio). *Preventing Pollution Through Technical Assistance: One State's Experience.* New York: 1990.

INFORM (Jacqueline B. Courteau and Nancy Lilienthal). *Toward A More Informed Public: Recommendations for Improving the Toxics Release Inventory.* New York: 1991.

Other related publications:

Congressional Office of Technology Assessment. *Serious Reduction of Hazardous Waste.* Washington, DC: September, 1986.

Congressional Office of Technology Assessment. *From Pollution to Prevention.* Washington, DC: June, 1987.

Hirschhorn, J. S. *Prosperity Without Pollution: The Prevention Strategy for Industry and Consumers.* New York: Van Nostrand Reinhold, 1991.

US Environmental Protection Agency. *Toxics in the Community: National and Local Perspectives: The 1989 Toxics Release Inventory National Report.* Washington DC: September, 1991.

US Environmental Protection Agency. *Pollution Prevention 1991: Progress on Reducing Industrial Pollutants.* Washington DC: October, 1991.

APPENDIX D

Glossary

Blowdown tank A tank that captures releases of vapors.

California list Waste regulated as hazardous waste in California in addition to the federal list of hazardous waste regulated under the Resource Conservation and Recovery Act (RCRA). The California list includes liquid waste containing certain metal ions, free cyanide, polychlorinated biphenyls, corrosives (pH less than 2.0), and liquids and non-liquids containing halogenated organics (i.e., organic compounds containing chlorine, bromine, iodine, or fluorine).

Catalyst A substance that increases or decreases the speed of a chemical reaction without undergoing a chemical change itself.

Chemical additive Substances used in product formulations to provide certain characteristics to the product. These characteristics might include color, elasticity, durability, viscosity, and others.

Chemical intermediate A chemical that when combined with a raw material initiates a reaction that leads to production of the final product. For example, peroxide initiates the free-radical polymerization reaction between vinyl chloride monomers.

Chemical substitution Replacement of hazardous chemicals with nonhazardous or less hazard-ous ones in both production and nonproduction processes.

Chemical-specific Applying to individual chemicals, versus broad classes of chemicals.

Cost accounting An account of all costs associated with the generation of a wastestream at the point at which it is generated. The account is done on a multimedia, chemical-specific basis and allows plant management to identify the contributions of each individual process to the plant's total waste generation. The following specific costs may be allocated to the individual process: materials costs (i.e., the costs of starting material and products lost); environmental handling costs (e.g., capital and operational expenses for treatment facilities, regulatory and compliance costs, waste transportation and disposal costs, and environmental liability insurance costs); insurance costs and future liabilities from hazardous wastes (e.g., from accidents, worker illness, or waste site cleanups). In addition to these specific costs, companies may also consider less quantifiable but important costs, such as those involved with public and customer relations related to waste problems.

Deming The William Deming philosophy of management. See *Total Quality Management (TQM)*.

Emergency Planning and Community Right-to-Know Act Title III of the Superfund Amendments and Reauthorization Act (SARA), passed by the US Congress in 1986: this major law gave the public significant new rights to find out about the dangerous chemicals stored, used, and released throughout the country. In particular, Section 313 of Title III created the Toxics Release Inventory (TRI) to provide public data on "routine" chemical releases from industries across the United States.

End-of-pipe The point at the end of the production process at which all products and waste products have been made and the waste products are being released (through a pipe, smokestack, or other release point); usually used as an adjective to refer to a pollution control strategy.

Equipment changes Modifications of and additions to equipment used in any stage of the manufacturing process (e.g., equipment used for storing, moving, mixing, or reacting chemicals) in order to reduce the amount of waste generated.

Fault tree analysis A schematic diagram of the flow of the individual parts of an industrial process indicating the types of problems that might occur, the possible consequences of these problems, and possible solutions.

Form R The form on which companies report Toxics Release Inventory data to state and federal environmental officials.

Fugitive air emissions Air pollution released through leaky valves, evaporation from tanks, and other *unintentional* release points.

Hazardous or toxic substances Defined by INFORM for the purposes of this study to mean materials included by the federal government on any of six lists, whether or not the substance is regulated in the medium in which it is released:

Clean Air Act hazardous air emissions

Clean Water Act priority pollutants

Emergency Planning and Community Right-to-Know Act (also known as Title III of the Superfund Amendments and Reauthorization Act, or SARA): Section 313 toxic substances; this is the list used by the EPA for the Toxics Release Inventory

Emergency Planning and Community Right-to-Know Act (also known as Title III of the Superfund Amendments and Reauthorization Act, or SARA): Section 302 extremely hazardous substances

Resource Conservation and Recovery Act (RCRA) Appendix VIII. hazardous constituents

Resource Conservation and Recovery Act (RCRA) commercial chemical products: U (hazardous wastes) and P (acute hazardous wastes)

Hazardous waste Under federal environmental law, refers specifically to solid hazardous discharges (not air pollutants) regulated under the federal Resource Conservation and Recovery Act (RCRA).

Heavy metal Metallic elements such as mercury, chromium, copper, zinc, lead, and cadmium having high molecular weights. These elements tend to be associated with negative health effects in humans above certain dose levels.

In-process recycling Recycling occurring as an integral part of the production process. For example, moving a waste stream from one end of an operation to a point near the front end in a closed-loop system for reuse as a raw material in that same operation.

Inorganic chemical In general, chemicals that do not contain the element carbon (the exceptions include certain simple carbon-containing compounds such as oxides [carbon monoxide, carbon dioxide], carbonates and bicarbonates [such as baking soda, baking powder, and chalk], cyanides and cyanates, and carbon disulfide). See *Organic chemical*.

Latex In the synthetic organic chemical industry, the term refers to any of various emulsions in water of a synthetic rubber or other polymer used chiefly in paint and other coatings and adhesives. Historically, latex also refers to the sap from a number of trees (including the rubber tree) from which natural rubber is produced.

Materials accounting A systematic tracking of raw materials and products as they move sequentially from one end of the plant to the other; not as quantitatively rigorous as a materials balance.

Materials balance A quantitative assessment of chemical inputs and outputs for individual processes that aims to account for every pound of a chemical that is (a) shipped to the process, (b) created or destroyed in the process, (c) delivered as a product from the process, and (d) wasted (irrespective of whether it is an air, water, or solid waste); if the amount of wastes identified does not equal the difference between the amount of the chemical entering (or being created in) and leaving (or being consumed in) the process, then other sources of waste must exist and need to be identified.

Materials Safety Data Sheet (MSDS) Part of the Hazard Communication Standards (HCS) set up by the US Occupational Safety and Health Administration (OSHA) to protect workers from chemical hazards. The MSDS provides the chemical composition of the substance being used, its trade name and name of the manufacturer, hazards associated with the substance, and precautions that workers should take to avoid such hazards.

Monomer A simple compound of low molecular weight capable of undergoing a chemical reaction in which it bonds with other such molecules to form very large compounds of very high molecular weight (polymers). Examples of common monomers are styrene, ethylene, acrylonitrile, vinyl chloride, and propylene.

MSDS See *Materials Safety Data Sheet*.

NPDES National Pollution Discharge Information System.

Multimedia Applying to all environmental media: land, water, and air.

Operational changes Changes in the way hazardous materials are handled at a plant (e.g., careful observation and control of materials, process conditions, and employee habits in order to minimize spills, process upsets, or the use of excessive amounts of chemicals) that can reduce generation of waste.

Organic chemical Chemical compounds containing carbon, except for certain simple ones. See *Inorganic chemical*.

Performance chemical Chemical substances designed to perform certain functions in the products they are contained in: for example, they allow certain products to withstand extremes in temperature, pressure, oxidation conditions, etc.

Plasticizer A chemical additive used in natural and synthetic polymers that imparts characteristics such as flexibility, elasticity, workability, color, etc.

Point source air emissions Air pollution released through smokestacks, vents, and other *intentional* release points.

Polymer Compounds of very high molecular weight made up of a large number of simple molecules (monomers) that have been caused to combine with each other through chemical reaction. Polymers can be naturally occurring, such as rubber, cellulose, starch, and proteins, or synthetic, such as polystyrene, nylon, polyethylene, and polypropylene.

POTWs See *Publicly owned treatment works*.

Priority Pollutants 126 specific chemicals regulated by the Clean Water Act amendments of 1977 as toxic chemicals. They include volatile substances, acidic, basic and neutral compounds, pesticides, metals, cyanides, and phenolic compounds.

Process changes Any change in the production process that reduces the generation of waste, ranging from simple alterations of process conditions such as temperature and pressure to discovery of new chemical pathways and production technologies.

Product changes   Changes in the product itself that can be achieved without changing the fundamental manufacturing process and that reduce the generation of waste (e.g., creating a chemical product in the form of pellets rather than as a powder can reduce the generation of waste dusts as the material is transferred during final packaging operations).

Publicly owned treatment works (POTWs)   Public sewage facilities.

Rare earth compounds   A group of 15 chemical elements (metals) having similar characteristics due to their electronic configurations; also called "lanthanides." The first in the series, lanthanum, is used in several alloys, as a catalyst, and in the glass industry.

RCRA   See *Resource Conservation and Recovery Act*.

Recycling   Reuse of by-products, or components of by-products, that might otherwise be disposed of in the environment.

Resin   A special category of polymers characterized by a tendency to harden upon heating (thermosetting), whereas other polymers soften (thermoplastic). Thermosetting polymers such as urea-formaldehyde and phenol-formaldehyde resins have a three-dimensional molecular structure and are the so-called "space-network polymers"; the molecular structure of thermoplastic polymers such as polyethylene and polyvinyl chloride are more two-dimensional and are thus called "linear or branched polymers."

Resource Conservation and Recovery Act (RCRA)   Federal "cradle to grave" regulations affecting hazardous and nonhazardous (garbage) solid waste.

Reuse   For this report, reuse refers to a substance that is re-introduced at the front end of a production process from which it was originally generated as a by-product.

Right-to-Know   A term usually referring to a series of laws, regulations, or databases that provide industry-related information to the public.

SARA   See *Superfund Amendments and Reauthorization Act*.

SIC codes   Standard Industrial Classification codes, the system the federal government uses to classify US companies according to the products they produce (e.g., the chemical and allied products industry is assigned SIC code 28, with individual industries in this category having four-digit codes that begin with 28).

Solvent   A substance, usually in liquid form, that serves as a medium in which other substances (solids, liquids, or gases) may be dissolved but does not react with those substances. The ability of solvents to dissolve other substances allows them to be used for cleaning purposes, as the major component of products such as paints and adhesives, or as the medium in which the dissolved chemicals may react with each other.

Source reduction activity (SRA)   An action or series of actions taken by plant management that avoids the creation of waste in the first place.

Source reduction   A strategy for reducing pollution that involves preventing the generation of waste in the first place rather than cleaning it up, treating it, or recycling after it has been produced.

Specialty chemical   Chemicals produced to serve particular functions in certain markets, such as providing protection against corrosion in automobiles.

SRA   See *Source reduction activity*.

Statistical process control (SPC) A set of systematic techniques to quantitatively confirm the relationship between process parameters and product quality. Examples of process parameters include temperature, pressure, reaction time, reagent concentrations, equipment size, equipment and instrumentation malfunctions, operator interaction, etc. SPCs can help to identify causes of process or product variations. SPC methods are applied to actual plant performance rather than at the process or product design stage.

Superfund Amendments and Reauthorization Act (SARA) A 1986 federal law amending the original "Superfund" law. Title III of this law is called the Emergency Planning and Community Right-to-Know Act (EPCRA). Section 313 of EPCRA contains the Toxics Release Inventory requirements.

Total Quality Management (TQM) A concept of business management, credited to William Deming, with the core principles of zero defects, statistical methods to measure success, and empowerment of workers to make improvements.

Toxic or hazardous substances See *Hazardous or toxic substances*.

Toxics Release Inventory (TRI) The US Environmental Protection Agency's annual inventory of the pounds of about 320 chemicals released to the air, water, or land or transferred off-site from the 20,000 or so largest manufacturing facilities using or manufacturing these chemicals in the United States. TRI provisions are found in Section 313 of the Emergency Planning and Community Right-to-Know Act, which is Title III of the 1986 Superfund Amendments and Reauthorization Act (SARA).

TRI See *Toxics Release Inventory*.

Vulcanization The process of heating a material in the presence of sulfur to impart certain characteristics.

Waste minimization Defined by EPA, in its 1986 report to Congress, as "the reduction, to the extent feasible, of hazardous waste that is generated or subsequently treated, stored, or disposed of. It includes any source reduction or recycling activity undertaken by a generator that results in either (1) the reduction of total volume or quantity of hazardous waste or (2) the reduction of toxicity of hazardous waste, or both, so long as such reduction is consistent with the goal of minimizing present and future threats to human health and the environment." It is concerned only with hazardous wastes as defined under the Resource Conservation and Recovery Act; this includes solid hazardous wastes and certain wastewaters destined for land disposal, but does not include air emissions.

Yield A measure of production efficiency. A common unit of measure is pounds of product produced per pound of raw material used, but other units of measure appropriate to the type of process used or product produced may also be used.

INDEX

A

Acids, 166, 168, 180, 183
    acidity (pH), 197
Additives, 128, 129, 136, 164, 167, 217, 224
Adhesives, 210-243
Air emissions, 16, 48, 50, 95, 125, 140, 147, 152, 191, 197, 219, 220, 231, 233, 235, 240, 243, 244
    nonattainment area, 228
    permits, 162
    process changes, 97
AMERICAN CYANAMID, 94-100
Antifoam agents, 141
Antitrust laws. see Regulations, antitrust
ARISTECH, 101-109
ATLANTIC INDUSTRIES, 110-114

B

Batch process, 34, 35, 133, 136, 202, 203, 234
Bayway Chemical Plant. see EXXON CHEMICAL AMERICAS
Biological oxygen demand (BOD), 97, 130, 132
Biological toxicity, 132
BOD. see Biological oxygen demand
BONNEAU DYE CORPORATION, 115
BORDEN CHEMICAL COMPANY, 116-122
By-products, 2, 155, 203, 230

C

California
    BORDEN CHEMICAL COMPANY, Fremont plant, 116-122
    CHEVRON CHEMICAL COMPANY, Richmond plant, 123-127
    COLLOIDS OF CALIFORNIA, Richmond plant, 141-143
    DOW CHEMICAL USA, Pittsburg plant, 147-163
    FIBREC, San Francisco plant, 171
    ICI AMERICAS, Richmond plant, 184-187
    ICI RESINS, Vallejo plant, 188-192
    SHELL CHEMICAL COMPANY, West Pittsburg plant, 240-242
    Tanner Act, 192
    UNOCAL CHEMICALS, La Mirada plant, 243-245
California Air Resources Board, 122
Capital
    expenditures, 96, 103, 242
    investment, 11, 12, 20, 21, 89, 103, 145, 202
        annual savings per dollar, 11, 12, 22-23
Capital costs. see Capital, investment
Carcinogens, 205
Carstab. see MORTON INTERNATIONAL
Catalysts, 166, 197, 240
Chemical intermediates. see Intermediate products
Chemical Manufacturers Association (CMA), 89, 235
    Community Awareness and Emergency Response (CAER) program, 108, 119, 125
    Responsible Care Program, 124, 230
Chemical substitution, 48, 49, 50, 90, 113, 123, 125, 197, 237

C (cont'd.)

Chemical substitution, definition, 9
CHEVRON CHEMICAL COMPANY, 123-127
CIBA-GEIGY CORPORATION, 128-140
Clean Air Acts, 2, 17, 228
Clean Water Acts, 123, 227
    priority pollutants, 2, 191
Closed plants, 128, 171, 182, 183
COLLOIDS OF CALIFORNIA, 141-143
Community/plant relations, 123, 162, 175
Computerization
    monitoring systems, 137, 163, 172, 244
    programs, 202, 203, 204
Containment, 2
Contamination, underground, 245
Continuous process, 34, 35
Cooperation, plant, with INFORM study, 7, 91, 123, 141, 144, 147, 154, 164, 188, 193, 194, 200, 202, 210, 217, 220, 229, 236
    non-cooperation, 115, 171, 183, 240
Corporate level offices, 155, 165, 205, 211, 221
Cosmetics, 193
Cost accounting, 31-32, 33, 90
    AMERICAN CYANAMID, 95
    ARISTECH, 103
    BORDEN CHEMICAL COMPANY, 117
    DOW CHEMICAL USA, 149
    DU PONT, 156
    EXXON CHEMICAL AMERICAS, 165, 166
    FISHER SCIENTIFIC COMPANY, 173
    MERCK AND COMPANY, 205
    MONSANTO COMPANY, 211
    PERSTORP POLYOLS, 221
    PMC SPECIALITIES GROUP, 225
Cost accounting, definition, 7, 8, 29
Costs
    allocated, 124, 231
    avoided, 32, 119
    capital. see Capital, investment
    operations, 3, 113, 145, 174, 214
    reduction programs, 89, 214, 215
    savings. see Savings, cost
    waste disposal, 45, 234

D

Data. see also Materials safety data sheets
    source reduction, 12-13, 88
    waste generation, collection by federal and state agencies, 217
Databases, 95, 148-149, 154, 156, 193, 211
DEF-TEC CORPORATION, 144-146
Detergents, 193
Distillation, 174, 180
DOW CHEMICAL USA, 147-163
DU PONT, 154-163
Dust, 139
    collectors, 136, 201
Dyes, 96-97, 110, 115, 128, 129, 133, 136, 154

E

E. I. Du Pont De Nemours and Company. see DU PONT
Economics
    chemical industry, 3
    efficiency, 172
    of source reduction
        benefits, 116, 155-156
        incentives, 228, 235, 244
Efficiency, 49, 174
    process, 203
    production, 89, 173
Electronics, 174
Emergency Planning and Community Right-to-Know Act. see Superfund Amendments and Reauthorization Act (SARA)
Emissions, air. see Air emissions
Employee involvement, 32, 33, 46, 90
    ARISTECH, 103
    BORDEN CHEMICAL COMPANY, 117
    DOW CHEMICAL USA, 149
    DU PONT, 154, 156, 158
    FISHER SCIENTIFIC COMPANY, 173
    ICI AMERICAS, 185
    ICI RESINS, 189
    MERCK AND COMPANY, 205
    MONSANTO COMPANY, 214
    PERSTORP POLYOLS, 221
    PMC SPECIALITIES GROUP, 225
    RHÔNE-POULENC, 231
Employee involvement, definition, 8, 29
Enamels, 188

E (cont'd.)

End-of-pipe controls, 159
Environmental Conservation and Recovery Act (ECRA) (New Jersey), 200
Environmental goals. see Goals, environmental
Environmental policy
    AMERICAN CYANAMID, 94-95
    ARISTECH, 102
    ATLANTIC INDUSTRIES, 110-111
    BORDEN CHEMICAL COMPANY, 116-117
    CHEVRON CHEMICAL COMPANY, 124
    CIBA-GEIGY CORPORATION, 129-130
    COLLOIDS OF CALIFORNIA, 141
    DEF-TEC CORPORATION, 144
    DOW CHEMICAL USA, 148
    DU PONT, 155-156
    ICI AMERICAS, 184-185
    ICI RESINS, 188-189
    INTERNATIONAL FLAVORS AND FRAGRANCES, 193
    MAX MARX COLOR AND CHEMICAL COMPANY, 200
    MERCK AND COMPANY, 202-203, 205
    MONSANTO COMPANY, 210-211
    MORTON INTERNATIONAL, 217
    PERSTORP POLYOLS, 220-221
    PMC SPECIALITIES GROUP, 224-225
    RHÔNE-POULENC, 229-230
    SHELL CHEMICAL COMPANY, 240
    UNOCAL CHEMICALS, 244
Environmental program, definition, 8, 29
Environmental programs, 33, 94-95, 149, 156, 166, 173, 214, 225, 231
Environmental Protection Agency (EPA)
    California, 125
    Ohio, 224
    US, 49, 193, 223
        Office of Toxic Substances, 112
Environmental regulations. see Regulations
Equipment, 23
    changes, 48, 50, 51, 90, 175, 207, 214, 234
    design, 180, 203
Equipment changes, definition, 9
Evaporation ponds, 150, 151
EXXON CHEMICAL AMERICAS, 164-170

F

FIBREC, 171
FISHER SCIENTIFIC COMPANY, 172-181
Flavor chemicals, 193
Fragrances, 193
France, regulatory structure, 229, 231-232
FRANK ENTERPRISES, 182
Fuel credits, 90, 211
Full cost accounting. see Cost accounting
Fungicides, 123

G

Gasoline additives, 123
Goals, environmental, 33, 94-95, 149, 156, 166, 205, 210-211, 213, 237
    definition, 8, 29
Government regulation. see Regulations
Groundwater, 102, 129

H

HART CHEM/J. E. HALMA, 183
Hazardous and Solid Waste Amendments (1984), 45
Hazardous chemicals, definition, 2
Hazardous waste, 1, 46, 183, 218, 225, 242
    amount and type reduced, 50
    definition, 2
Hazardous Waste Generator Waste Minimization Report (New Jersey), 193
Hazardous Waste Source Reduction and Management Review Act of 1989, 126
Heavy metals, 132-133, 201
Herbicides, 123, 184
Housekeeping, 141, 151, 201, 221

I

ICI AMERICAS, 184-187
ICI RESINS, 188-192
Imperial Chemical Industries. see ICI AMERICAS; ICI RESINS
Implementation time, 13, 17-18, 25, 90
Incentives, for source reduction. see Motivation for source reduction; Regulations, effect on motivation of source reduction
Incineration, 46, 168, 215, 234

I (cont'd.)

INFORM
    scope of study, 10
    study methodology, 6-10
        questionaire, 7
    study plants. see Plant profiles
Inorganic chemicals, 147, 200
Intermediate products, 152, 158, 200, 224, 229, 236
INTERNATIONAL FLAVORS AND FRAGRANCES, 195-199
Inventory control, 190, 236, 240

L

Lacquers, 188
Land disposal, 165, 168, 230, 234
Landfills, 94, 95, 151, 159, 162
Latexes, 147, 151, 210, 243
    polymer, 243
Leadership, 32, 90
    AMERICAN CYANAMID, 95
    ARISTECH, 103
    ATLANTIC INDUSTRIES, 111
    DOW CHEMICAL USA, 149
    DU PONT, 156
    FISHER SCIENTIFIC COMPANY, 32, 173
    MONSANTO COMPANY, 214
    PERSTORP POLYOLS, 221
    PMC SPECIALTIES GROUP, 225
    RHÔNE-POULENC, 231
    SCHER CHEMICALS, 237

Leadership, definition, 8, 29
Legislation. see Regulations
Liability, 46
Litigation, 128
Lubricants. see Oils
Lubricating oil additives. see Additives

M

Maintenance operations, 23
Management
    environmental, 90, 148
    production managers, 32, 205
    responsibility for source reduction progress, 7, 96, 111, 155, 205

Mass balance accounting, 165
Materials accounting, 33
    AMERICAN CYANAMID, 95
    DOW CHEMICAL USA, 149
    DU PONT, 156
    EXXON CHEMICAL AMERICAS, 166
    FISHER SCIENTIFIC COMPANY, 173
    MERCK AND COMPANY, 205
    MONSANTO COMPANY, 211
    PMC SPECIALTIES GROUP, 225
    RHÔNE-POULENC, 231-232
    SCHER CHEMICALS, 237
Materials accounting, definition, 8, 29
Materials balance, 33, 124, 132, 149, 173, 189, 203
    definition, 8, 29
Materials data collection
    AMERICAN CYANAMID, 95
    ARISTECH, 102-103
    BORDEN CHEMICAL COMPANY, 117
    CHEVRON CHEMICAL COMPANY, 124
    CIBA-GEIGY CORPORATION, 130
    DEF-TEC CORPORATION, 145
    DOW CHEMICAL USA, 148-149
    DU PONT, 156
    EXXON CHEMICAL AMERICAS, 165
    FISHER SCIENTIFIC COMPANY, 172-173
    ICI AMERICAS, 185
    ICI RESINS, 189
    INTERNATIONAL FLAVORS AND FRAGRANCES, 193-194
    MAX MARX COLOR AND CHEMICAL COMPANY, 200-201
    MERCK AND COMPANY, 203-204
    MONSANTO COMPANY, 211
    MORTON INTERNATIONAL, 217, 219
    PERSTORP POLYOLS, 221
    PMC SPECIALTIES GROUP, 225
    RHÔNE-POULENC, 230
    SCHER CHEMICALS, 236
    SHELL CHEMICAL COMPANY, 240
    UNOCAL CHEMICALS, 244
Materials safety data sheets (MSDS), 141, 162
Materials tracking system, 7, 94, 162, 172-173, 181
MAX MARX COLOR AND CHEMICAL COMPANY, 200-201

M (cont'd.)

Medium of waste reduced, 16-17. see also Waste management practices
MERCK AND COMPANY, 202-207
    Toxics Release Inventory, 206
Monomers, 211, 215
MONSANTO COMPANY, 210-216
MORTON INTERNATIONAL, 217-219
Motivation for source reduction activities, 44-48
    factors data, 13

N

National Pollutant Discharge Elimination System (NPDES), 102
National Priority List. see Superfund amendments
New Jersey
    ATLANTIC, Nutley plant, 110-114
    CIBA-GEIGY CORPORATION, Toms River plant, 128-140
    DU PONT, Deepwater plant, 154-163
    EXXON CHEMICALS AMERICA, Linden plant, 164-170
    FISHER SCIENTIFIC COMPANY, Fair Lawn plant, 172-181
    HART CHEM/J. E. HALMA, Garfield plant, 183
    Hazardous Waste Generator Waste Minimization Reports, 193, 197
    Hazardous Waste Minimization Survey, 6, 172, 207
    INTERNATIONAL FLAVORS AND FRAGRANCES, Union Beach plant, 195-199
    MAX MARX COLOR AND CHEMICAL COMPANY, Irvington plant, 200-201
    MERCK AND COMPANY, Rahway plant, 202-207
    RHÔNE-POULENC, New Brunswick plant, 229-235
    Right-to-Know report forms, 6
    SCHER CHEMICALS, Clifton plant, 236-239
    state regulations, 183
New Jersey Department of Environmental Protection, 128, 183
Nonproduction functions. see Source reduction, nonproduction functions

O

Office of Technology Assessment (OTA), 2
Off-site waste disposal, 234
    recovery, 207
    transfers, 194, 240
Off-specification products, 2
Ohio
    AMERICAN CYANAMID, Marietta plant, 94-100
    ARISTECH, Haverhill plant, 101-109
    BONNEAU DYE CORPORATION, Avon plant, 115
    DEF-TEC CORPORATION, Rock Creek plant, 144-146
    FRANK ENTERPRISES, Columbus plant, 182
    MONSANTO COMPANY, Addyston plant, 210-216
    MORTON INTERNATIONAL, Cincinnati plant, 217-219
    PERSTORP POLYOLS, Toledo plant, 220-223
    PMC SPECIALITIES GROUP, Cincinnati plant, 224-228
Ohio EPA Air Agency, 146
Ohio Manufacturers Association, 100
Oils, 129, 162, 166-167, 217, 220, 236, 240
Operations changes, 25, 48, 50, 51, 90, 175, 197, 214-215
    definition, 9
Operations managers. see Management, production managers
Operator handling, 204
Organic chemicals, 147, 200, 202, 224, 243
Ownership, changes in, 35

P

Paints, 210, 220, 243
Paper industry, 129
Payback periods, 2, 11, 12, 21, 22, 25
Performance chemicals, 217
PERSTORP POLYOLS, 220-223
Pesticides, 123, 202
Pharmaceuticals, 129, 137
Pigments, 200
Plant activity, changes in, 7
Plant cooperation. see Cooperation, plant
Plant profiles, 4-5, 6-7, 9, 10, 91-245
    components of, 91-92

P (cont'd.)

Plastics, 128, 129, 136, 210, 217, 220
PMC SPECIALITIES GROUP, 224-228
Pollution control
    costs, 1
    versus source reduction, 2
Pollution Prevention Act (New Jersey), 1991, 45, 181
Pollution Prevention Act (US) (1991), 3
Polymers, 188, 191, 210, 215, 243
Polyvinyl Chemical industries. see ICI RESINS
Port Plastics Plant. see MONSANTO
Process changes, 7, 35, 48, 50, 51, 90, 132-133, 166, 174, 180, 203, 214, 219, 225
    definition, 9
Process types. see Batch process; Continuous process
Product changes, definition, 9
Product development. see Research and development
Production efficiency. see Efficiency, production
Production functions. see Source reduction, production functions
Products
    changes, 48, 50, 51, 90, 97
    output, 46
    storage, 23
Products and operations
    AMERICAN CYANAMID, 94
    ARISTECH, 101-012
    ATLANTIC INDUSTRIES, 110
    BORDEN CHEMICAL COMPANY, 116
    CHEVRON CHEMICAL COMPANY, 123
    COLLOIDS OF CALIFORNIA, 141
    DEF-TEC CORPORATION, 144
    DOW CHEMICAL USA, 147-148
    DU PONT, 154-155
    FISHER SCIENTIFIC COMPANY, 172
    ICI AMERICAS, 184
    ICI RESINS, 188
    INTERNATIONAL FLAVORS AND FRAGRANCES, 193
    MAX MARX COLOR AND CHEMICAL COMPANY, 200
    MERCK AND COMPANY, 202
    MONSANTO COMPANY, 210
    PERSTORP POLYOLS, 220
    PMC SPECIALITIES GROUP, 224
    RHÔNE-POULENC, 229
    SCHER CHEMICALS, 236
    SHELL CHEMICAL COMPANY, 240
    UNOCAL CHEMICALS, 243
Product yields. see Yields, product
Profiles. see Plant profiles
Proprietary data, 123, 187, 235

Q

Quality assurance sampling, 125
Quality control programs
    CIBA-GEIGY CORPORATION, 131
    DU PONT, 158
    FISHER SCIENTIFIC COMPANY, 175
    ICI RESINS, 189
    MONSANTO COMPANY, 214
    SHELL CHEMICAL COMPANY, 242

R

Raw materials, 23, 123, 147, 158, 174, 190, 203, 219, 231, 234
    storage, 23
RCRA. see Resource Conservation and Recovery Act
Reclamation. see Recycling
Recycling, 151, 154, 155, 164, 167, 174, 175, 203, 219, 229, 231, 236, 237, 240
    closed loop, 214, 216, 225, 226, 230
Regulations
    air permits, 228
    antitrust, 235
    Clean Water Acts, 123
    effect on motivation of source reduction, 9, 45-46, 90, 94, 139, 147, 156, 230
    Environmental Conservation and Recovery Act (ECRA) (New Jersey), 200
    federal, 3, 96, 164, 230
    mandatory, 216
    National Pollutant Discharge Elimination System (NPDES), 102
    ocean discharge permits, 128
    permits, 102, 162
    plant perspectives on, 100, 108-109, 114, 139, 143, 146, 151, 162-163, 181, 192, 197, 216, 222-223, 228, 235, 237, 245

R (cont'd.)

Regulations (continued)
- Resource Conservation and Recovery Act (RCRA). see Resource Conservation and Recovery Act (RCRA)
- source reduction. see Resource Conservation and Recovery Act (RCRA)
- state, 116, 230
  - California, 124, 147, 244
    - Hazardous Waste Source Reduction and Management Review Act of 1989, 126
    - State Proposition 65, 244
    - Tanner Act, 192
  - New Jersey, 172, 237
    - Environmental Conservation and Recovery Act (ECRA), 200
    - Pollution Prevention Act, 1991, 181
  - Ohio, 6, 96, 221
- Research and development, 90, 97, 130, 169, 202, 225-226, 231
- Resins, 128, 129, 188, 210, 243
- Resistance, corrosion, 129
- Resource Conservation and Recovery Act (RCRA), 2, 45, 89, 94, 96, 102, 108, 124, 147, 155, 173, 181, 193, 194, 201, 219, 224, 240, 242
- Responsible Care Program. see Chemical Manufacturers Association (CMA), Responsible Care Program
- RHÔNE-POULENC, 229-235

S

Safety
- environmental, 232
- public health, 232
- worker. see Employee involvement

Salvage. see Recycling
Savings, cost, 19-20, 22-23, 32, 234
SCHER CHEMICALS, 236-239
Semiconductor industry, 183
Sewer systems, sewage treatment, 97
SHELL CHEMICAL COMPANY, 240-242
Sherman Williams. see PMC SPECIALITIES GROUP
Size of plants, significance of, 33, 35
Soil disposal, 215, 226
Solid waste, 16, 45, 48, 51, 94, 95, 155-156, 165, 233
- management plan, EXXON, 162, 165

Solvents, 23, 49, 173, 174, 175, 180, 207, 225, 243
Source reduction
- accomplishments at the study plants, 3, 14-26, 30
- activities
  - AMERICAN CYANAMID, 96-97
  - ARISTECH, 103-107
  - ATLANTIC INDUSTRIES, 111-114
  - BORDEN CHEMICAL COMPANY, 117-118, 120-121
  - CHEVRON CHEMICAL COMPANY, 125-126, 127
  - CIBA-GEIGY CORPORATION, 132-136
  - COLLOIDS OF CALIFORNIA, 142
  - DEF-TEC CORPORATION, 144, 145
  - DOW CHEMICAL USA, 150-153
  - DU PONT, 158-161
  - EXXON CHEMICAL AMERICAS, 169
  - FIBREC, 174-180
  - ICI AMERICAS, 185-187
  - ICI RESINS, 189-191
  - INTERNATIONAL FLAVORS AND FRAGRANCES, 196-197
  - MERCK AND COMPANY, 207, 208-209
  - MONSANTO COMPANY, 212-213, 214-215
  - PERSTORP POLYOLS, 221
  - PMC SPECIALITIES GROUP, 225-228
  - RHÔNE-POULENC, 232, 233-234
  - SCHER CHEMICALS, 237
  - SHELL CHEMICAL COMPANY, 242
  - UNOCAL CHEMICALS, 244
- changes in activity, 7, 96-97, 196
- data from 1978-88, 3, 11-13
- economic benefits, 3, 12, 116, 155-156
- environmental benefits, 2
- implementation mechanisms, 7
- nonproduction functions, 23-25, 26
- obstacles to, 244
- production functions, 24, 26
- program features, 30-33, 36-43, 90
- research on, 6, 7, 203

S (cont'd.)

Source reduction (continued)
    techniques, 8-9, 13, 48, 49, 51, 52-87
    written policies, 7, 8, 33, 202-203, 217, 220-221, 224, 229, 231, 240

Source reduction, definitions, 8-9

Speciality chemicals, 129, 164, 224, 236

Spills, 7, 189, 190, 215, 221, 224, 231

Statistical data tests, 10

Statistical tracking and analysis, statistical process controls (SPC), 96, 117, 124, 215

Stauffer. see ICI AMERICAS

Study plants. see Plant profiles

Substitutions. see Chemical substitutions

Superfund Amendments and Reauthorization Act (SARA), 48-49, 140, 219
    National Priority List, 128
    section 313, 2, 108, 173, 240
    section 302 extremely hazardous substances, 2

Sweden, environmental regulations, 223

T

Technical assistance, 9
    AMERICAN CYANAMID, 99-100
    ARISTECH, 108
    ATLANTIC INDUSTRIES, 114
    BORDEN CHEMICAL COMPANY, 119, 122
    CIBA-GEIGY CORPORATION, 139
    COLLOIDS OF CALIFORNIA, 143
    DEF-TEC CORPORATION, 146
    DOW CHEMICAL USA, 151, 237
    DU PONT, 162, 174
    EXXON CHEMICAL AMERICAS, 169-170
    FISHER SCIENTIFIC COMPANY, 174, 181
    ICI AMERICAS, 185, 187
    ICI RESINS, 192
    INTERNATIONAL FLAVORS AND FRAGRANCES, 197
    MAX MARX COLOR AND CHEMICAL COMPANY, 201
    MONSANTO COMPANY, 215-216
    PERSTORP POLYOLS, 222
    PMC SPECIALITIES GROUP, 228
    RHÔNE-POULENC, 235
    SCHER CHEMICALS, 237
    SHELL CHEMICAL COMPANY, 242

Techniques, source reduction. see Source reduction, techniques

Textile industry, 129

Toms River Plant. see CIBA-GEIGY

Total organic carbon (TOC), 132, 136, 162, 197

Toxic and hazardous wastes, definition, 2

Toxics Release Inventory (TRI), 1, 16, 108-109, 123, 194, 206, 211, 219
    releases and transfers, 11, 88-90, 195, 205, 218, 230, 240, 241

Toxic Substances Control Act (TSCA), 102, 152

Tracking process, 124

Trade associations, 235

Trade secrets. see Proprietary data

Training programs. see Employee involvement

Type of waste reduced, 25, 47-48, 51. see also Solid waste; Waste management practices; Wastestream; Wastewater

U

UNOCAL CHEMICALS, 243-245

USS Chemicals. see ARISTECH

V

Volatile organic compounds (VOC), 221, 231
    emissions, 100, 191

Volatile organic substances (VOS), 137, 234, 235

W

Waste disposal, costs. see Costs, waste disposal

Waste index, 149

Waste management practices
    AMERICAN CYANAMID, 98-99
    ARISTECH, 107-108
    ATLANTIC INDUSTRIES, 114
    CIBA-GEIGY CORPORATION, 128, 137-139
    DEF-TEC CORPORATION, 146
    FISHER SCIENTIFIC COMPANY, 180-181
    ICI RESINS, 191
    MAX MARX COLOR AND CHEMICAL COMPANY, 201
    MONSANTO COMPANY, 215
    MORTON INTERNATIONAL, 219

W (cont'd.)

Waste management practices (continued)
    PERSTORP POLYOLS, 221-222
    PMC SPECIALITIES GROUP, 228
    RHÔNE-POULENC, 234
    UNOCAL CHEMICALS, 244-245

Waste minimization, 50, 103, 114, 154, 156, 165, 172, 173, 184, 187, 193, 196, 203, 207, 214, 229-230, 242
    percentage reduction, 14, 15

Waste reduction. see Waste minimization

Wastestream, 2, 15, 22-23, 103, 125, 165, 197, 202, 204
    average percent reduction in, 12-13, 14-15, 50
    generation (table), 204
    reduction, 90, 197

Wastewater, 16, 48, 51, 95, 114, 118-119, 122, 128, 132, 150, 154, 156, 159, 191, 200, 215, 220, 221, 222, 225, 228, 234
    effluent reductions, 138, 200, 216
    pretreatment, 231

Worker safety. see Employee involvement

Written source reduction policy, 33
    definition, 8, 29

Y

Yields
    average percent increase, 13, 18, 19
    product, 90, 155, 173, 174, 189, 197, 214, 215, 216, 227, 230, 231, 234

ABOUT THE AUTHORS

Mark H. Dorfman

Mark Dorfman joined INFORM's Chemical Hazards Prevention Program in 1987 as Research Associate. He now holds the position of Associate Director of that program. He is the principal author of *Pollution Prevention Through Technical Assistance: One State's Experience*. He has conducted pollution prevention workshops for citizen groups around the United States and in several cities internationally.

Prior to joining INFORM, Mr. Dorfman worked as a consultant for UNICEF/Nepal. He conducted an industrial hazardous waste survey in Nepal's Katmandu Valley for the West German Technical Assistance Agency, and worked as an analytical chemist in the United States.

Mr. Dorfman received a B.S. from Ramapo College of New Jersey and an M.S. in environmental chemistry from the University of North Carolina's School of Public Health at Chapel Hill.

Catherine G. Miller, Ph.D.

Dr. Catherine Miller has been a consultant to INFORM since 1984. She also works as a consultant to other organizations on projects analyzing the Toxics Release Inventory and other toxic data resources for federal and state agencies. Dr. Miller was a coauthor of INFORM's study, *Cutting Chemical Wastes*.

Dr. Miller also worked for the Environmental Protection Agency from 1971 to 1975 and 1978 to 1984, and is the author of half a dozen EPA studies, including assessments of the economic benefits of municipal wastewater treatment and multimedia pollution control programs and the policy uses of air quality models and exposure assessment modeling. As an EPA operations research analyst, she developed management and budget analyses of operational and policy issues, including data information systems.

Dr. Miller holds a B.A. from Smith College, an M.S. from the Massachusetts Institute of Technology, and a Ph.D. in public policy from Harvard University.

Warren R. Muir, Ph.D.

Dr. Warren Muir has been a key advisor to INFORM's Chemical Hazards Prevention Program since June 1982, and is Senior Fellow of that program. He is the coauthor of two INFORM studies: *Cutting Chemical Wastes* and *Promoting Hazardous Waste Reduction: Six Steps States Can Take*. Dr. Muir is founder and president of Hampshire Research Institute, an environmental and scientific information firm. He is frequently invited to testify on pollution prevention–related federal and state legislation.

Prior to becoming affiliated with INFORM as a consultant, Dr. Muir served with the US Environmental Protection Agency as Director of the Office of Toxic Substances in 1980-1981, and with the President's Council on Environmental Quality as Senior Staff Member from 1971 to 1978. Dr. Muir has served as a member of or advisor to many national and international environmental organizations and government agencies, including the National Academy of Sciences, and is on the faculty of the Johns Hopkins School of Hygiene and Public Health.

Dr. Muir earned his B.A. from Amherst College, and holds an M.S. and Ph.D. in chemistry form Northwestern University.

INFORM PUBLICATIONS AND MEMBERSHIP

Selected Publications on Chemical Hazards Prevention

Tackling Toxics in Everyday Products: A Directory of Organizations (Nancy Lilienthal, Michèle Ascione, Adam Flint), 1992, 180 pp., $19.95.

Toward A More Informed Public: Recommendations for Improving the Toxics Release Inventory (Jaqueline B. Courteau and Nancy Lilienthal), 1991, 24 pp., $15.00.

Toxic Clusters: Patterns of Pollution in the Midwest (Nancy Lilienthal and Sibyl R. Golden), 1991, 108 pp., $15.00.

A Citizen's Guide to Promoting Toxic Waste Reduction (Lauren Kenworthy), 1990, 128 pp., $15.00. (Revised edition to be published in 1992.)

Trading Toxics Across State Lines (Nancy Lilienthal), 1990, 32 pp., $7.50.

Preventing Pollution Through Technical Assistance: One State's Experience (Mark Dorfman and John Riggio), 1990, 72 pp., $15.00.

Cutting Chemical Wastes: What 29 Organic Chemical Plants Are Doing to Reduce Hazardous Wastes (David J. Sarokin, Warren Muir, Ph.D., Catherine G. Miller, Ph.D., and Sebastian R. Sperber), 1985, 548 pp., $47.00.

Other INFORM Publications

INFORM also publishes reports on municipal solid waste management, urban air quality, and land and water conservation, and a quarterly newsletter. For a complete publications list or more information, call or write to INFORM.

Sales Information

Payment

Payment, including shipping and handling charges, must be in US funds drawn on a US bank and must accompany all orders. Please make checks payable to INFORM and mail to:

> INFORM
> 381 Park Avenue South
> New York, NY 10016-8806

Please include a street address; UPS cannot deliver to a box number.

Shipping Fees

To order in the US, please send a check that includes $3.00 for the first book and $1.00 for each additional book for shipping and handling charges. To order in Canada, add $5.00 for the first book and $3.00 for each additional book. For information on shipping rates for other countries, call (212) 689-4040.

Discount Policy

Booksellers:	20% on 1-4 copies of same title
	30% on 5 or more copies of same title
General bulk:	20% on 5 or more copies of same title

Public interest and community groups:

	Price:
Books under $10:	No discount
Books $10-$25:	$10
Books $25 and up:	$15

Returns

Booksellers may return books, if in saleable condition, for full credit or cash refund up to 6 months from date of invoice. Books must be returned prepaid and include a copy of the invoice or packing list showing invoice number, date, list price, and original discount.

Membership

Individuals provide an important source of support to INFORM and receive the following benefits:

Member ($25):	A one-year subscription to *INFORM Reports*, INFORM's quarterly newsletter, and early notice of new publications.
Friend ($50):	Member's benefits and a string bag.
Contributor ($100):	Friend's benefits, plus a 10% discount on new INFORM studies.
Supporter ($250):	Friend's benefits, plus a 20% discount on new INFORM studies.
Donor ($500):	Friend's benefits, plus a 30% discount on new INFORM studies.
Associate ($1000):	Friend's benefits, plus a complimentary copy of new INFORM studies.
Benefactor ($5000):	Friend's benefits, plus a complimentary copy of new INFORM studies.

INFORM BOARD OF DIRECTORS

Charles A. Moran, Chair
President
Government Securities Clearing Corporation

Kiku Hoagland Hanes, Vice Chair
Vice President
The Conservation Fund

James B. Adler
President
Adler & Adler Publishers

Victor G. Alicea
President
Boricua College

Paul A. Brooke
Managing Director
Morgan Stanley & Co., Inc.

Christopher J. Daggett
Managing Director
William E. Simon & Sons, Inc.

Michael J. Feeley
President and Chief Executive Officer
Feeley & Willcox

Barbara D. Fiorito
Vice President, Marketing
& Communications
Spears Benzak Salomon & Farrell

Jane R. Fitzgibbon
Senior Vice President
Group Director
Ogilvy & Mather Advertising

C. Howard Hardesty, Jr.
Partner
Andrews & Kurth

Lawrence S. Huntington
Chairman of the Board
Fiduciary Trust Company International

Martin Krasney
President
Center for the Twenty-First Century

Philip J. Landrigan, M.D.
Ethel H. Wise Professor of Community Medicine
Chairman, Department of Community Medicine
The Mount Sinai Medical Center

Dr. Jay T. Last
President
Hillcrest Press

Joseph T. McLaughlin
Partner
Shearman & Sterling

Kenneth F. Mountcastle, Jr.
Senior Vice President
Dean Witter Reynolds, Inc.

Susan Reichman
Communications and Marketing Consultant

S. Bruce Smart, Jr.
Senior Counselor
World Resources Institute

Frank T. Thoelen
Partner
Arthur Andersen & Co.

Grant P. Thompson
Executive Vice President
The Wilderness Society

Joanna D. Underwood
President
INFORM, Inc.